Lecture Notes in Mathematics 1905

Editors:
J.-M. Morel, Cachan
F. Takens, Groningen
B. Teissier, Paris

Claudia Prévôt · Michael Röckner

A Concise Course on Stochastic Partial Differential Equations

Authors

Claudia Prévôt
Fakultät für Mathematik
Universität Bielefeld
Universitätsstr. 25
33615 Bielefeld
Germany
e-mail: cprevot@web.de

Michael Röckner
Fakultät für Mathematik
Universität Bielefeld
Universitätsstr. 25
33615 Bielefeld
Germany
e-mail: roeckner@math.uni-bielefeld.de

Departments of Mathematics
  and Statistics
Purdue University
150 N. University St.
West Lafayette, IN 47907-2067
USA
e-mail: roeckner@math.purdue.edu

Library of Congress Control Number: 2007925694

Mathematics Subject Classification (2000): 35-XX, 60-XX

ISSN print edition: 0075-8434
ISSN electronic edition: 1617-9692
ISBN-10   3-540-70780-8 Springer Berlin Heidelberg New York
ISBN-13   978-3-540-70780-6 Springer Berlin Heidelberg New York

DOI 10.1007/978-3-540-70781-3

Springer is a part of Springer Science+Business Media
springer.com
© Springer-Verlag Berlin Heidelberg 2007

Typesetting by the authors and SPi using a Springer LATEX macro package
Cover design: WMXDesign GmbH, Heidelberg

Printed on acid-free paper   SPIN: 11982159   VA41/3100/SPi   5 4 3 2 1 0

Contents

Bibliography   137

Index   140

Symbols   143

1. Motivation, Aims and Examples

These lectures will concentrate on (nonlinear) stochastic partial differential equations (SPDEs) of evolutionary type. All kinds of dynamics with stochastic influence in nature or man-made complex systems can be modelled by such equations. As we shall see from the examples, at the end of this section the state spaces of their solutions are necessarily infinite dimensional such as spaces of (generalized) functions. In these notes the state spaces, denoted by E, will be mostly separable Hilbert spaces, sometimes separable Banach spaces.

There is also enormous research activity on SPDEs, where the state spaces are not linear, but rather spaces of measures (particle systems, dynamics in population genetics) or infinite-dimensional manifolds (path or loop spaces over Riemannian manifolds).

There are basically three approaches to analysing SPDEs: the "martingale (or martingale measure) approach" (cf. [Wal86]), the "semigroup (or mild solution) approach" (cf. [DPZ92], [DPZ96]) and the "variational approach" (cf. [Roz90]). There is an enormously rich literature on all three approaches which cannot be listed here. We refer instead to the above monographs.

The purpose of these notes is to give a concise introduction to the "variational approach", as self-contained as possible. This approach was initiated in pioneering work by Pardoux ([Par72],[Par75]) and further developed by N. Krylov and B. Rozowskii in [KR79] (see also [Roz90]) for continuous martingales as integrators in the noise term and later by I. Gyongy and N. Krylov in [GK81],[GK82],[Gyö82] for not necessarily continuous martingales.

These notes grew out of a two-semester graduate course given by the second-named author at Purdue University in 2005/2006. The material has been streamlined and could be covered in just one semester depending on the pre-knowledge of the attending students. Prerequisites would be an advanced course in probability theory, covering standard martingale theory, stochastic processes in $\mathbb{R}^d$ and maybe basic stochastic integration, though the latter is not formally required. Since graduate students in probability theory are usually not familiar with the theory of Hilbert spaces or basic linear operator theory, all required material from these areas is included in the notes, most of it in the appendices. For the same reason we minimize the general theory of martingales on Hilbert spaces, paying, however, the price that some proofs about stochastic integration on Hilbert space are a bit lengthy, since they have to be done "by bare hands".

In comparison with [Roz90] for simplicity we specialize to the case where the integrator in the noise term is just a cylindrical Wiener process. But everything is spelt out in a way so that it generalizes directly to continuous local martingales. In particular, integrands are always assumed to be predictable rather than just adapted and product measurable. The existence and uniqueness proof (cf. Subsection 4.2) is our personal version of the one in [KR79], [Roz90] and largely taken from [RRW06] presented there in a more general framework. The results on invariant measures (cf. Subsection 4.3) we could not find in the literature for the "variational approach". They are, however, quite straightforward modifications of those in the "semigroup approach" in [DPZ96]. The examples and applications in Subsection 4.1 in connection with the stochastic porous media equation are fairly recent and are modifications from results in [DPRLRW06] and [RRW06].

To keep these notes reasonably self-contained we also include a complete proof of the finite-dimensional case in Chapter 3, which is based on the very focussed and beautiful exposition in [Kry99], which uses the Euler approximation. Among other complementing topics the appendices contain a detailed account of the Yamada–Watanabe theorem on the relation between weak and strong solutions (cf. Appendix E).

The structure of these notes is, as we hope, obvious from the list of contents. We only would like to mention here, that a substantial part consists of a very detailed introduction to stochastic integration on Hilbert spaces (see Chapter 2), major parts of which (as well as Appendices A–C) are taken from the Diploma thesis of Claudia Prévôt and Katja Frieler. We would like to thank Katja Frieler at this point for her permission to do this. We also like to thank all coauthors of those joint papers which form another component for the basis of these notes. It was really a pleasure working with them in this exciting area of probability. We would also like to thank Matthias Stephan and Sven Wiesinger for the excellent typing job, as well as the participants of the graduate course at Purdue University for spotting many misprints and small mistakes.

Before starting with the main body of these notes we would like to give a few examples of SPDE that appear in fundamental applications. We do this in a very brief way, in particular, pointing out which of them can be analysed by the tools developed in this course. We refer to the above-mentioned literature for a more elaborate discussion of these and many more examples and their role in the applied sciences.

Example 1.0.1 (Stochastic quantization of the free Euclidean quantum field).

$$\mathrm{d}X_t = (\Delta - m^2)X_t \, \mathrm{d}t + \mathrm{d}W_t$$

on $E \subset \mathcal{S}'(\mathbb{R}^d)$.

- $m \in [0, \infty)$ denotes "mass",

- $(W_t)_{t \geqslant 0}$ is a cylindrical Brownian motion on $L^2(\mathbb{R}^d) \subset E$ (the inclusion is a Hilbert–Schmidt embedding).

Example 1.0.2 (Stochastic reaction diffusion equations).

$$\mathrm{d}X_t = [\Delta X_t - X_t^3] \, \mathrm{d}t + \sqrt{Q} \, \mathrm{d}W_t$$

on $E := L^p(\mathbb{R}^d)$.

- Q is a trace class operator on $L^2(\mathbb{R}^d)$, can also depend on X_t (then Q becomes $Q(X_t)$),

- $(W_t)_{t \geqslant 0}$ is a cylindrical Brownian motion on $L^2(\mathbb{R}^d)$.

Example 1.0.3 (Stochastic Burgers equation).

$$\mathrm{d}X_t = \Delta X_t - X_t \frac{\mathrm{d}}{\mathrm{d}\xi} X_t + \sqrt{Q} \, \mathrm{d}W_t$$

on $E := L^2([0, 1])$.

- $\xi \in [0, 1]$,

- Q as above,

- $(W_t)_{t \geqslant 0}$ is a cylindrical Brownian motion on $L^2([0, 1])$.

Example 1.0.4 (Stochastic Navier–Stokes equation).

$$\mathrm{d}X_t = [\nu \Delta_s X_t - \langle X_t, \nabla \rangle X_t] \, \mathrm{d}t + \sqrt{Q} \, \mathrm{d}W_t$$

on $E := \{x \in L^2(\Lambda \to \mathbb{R}^2, \, \mathrm{d}x) \mid \mathrm{div}\, x = 0\}$, $\Lambda \subset \mathbb{R}^d$, $d = 2, 3$, $\partial\Lambda$ smooth.

- ν denotes viscosity,

- Δ_s denotes the Stokes Laplacian,

- Q as above,

- $(W_t)_{t \geqslant 0}$ is a cylindrical Brownian motion on $L^2(\Lambda \to \mathbb{R}^d)$,

- div is taken in the sense of distributions.

Example 1.0.5 (Stochastic porous media equation).

$$\mathrm{d}X_t = [\Delta \Psi(X_t) + \Phi(X_t)] \, \mathrm{d}t + B(X_t) \, \mathrm{d}W_t$$

on $H := $ dual of $H_0^1(\Lambda)$ (:= Sobolev space of order 1 in $L^2(\Lambda)$ with Dirichlet boundary conditions).

- Λ as above,

- $\Psi, \Phi : \mathbb{R} \to \mathbb{R}$ "monotone",

- $B(x) : H \to H$ Hilbert–Schmidt operator, $\forall\, x \in H$.

The general form of these equations with state spaces consisting of functions $\xi \mapsto x(\xi)$, where ξ is a spatial variable, e.g. from a subset of $\mathbb{R}^d$, looks as follows:

$$\mathrm{d}X_t(\xi) = A\Big(t, X_t(\xi), D_\xi X_t(\xi), D_\xi^2\big(X_t(\xi)\big)\Big)\, \mathrm{d}t$$

$$+ B\Big(t, X_t(\xi), D_\xi X_t(\xi), D_\xi^2\big(X_t(\xi)\big)\Big)\, \mathrm{d}W_t .$$

Here D_ξ and D_ξ^2 mean first and second total derivatives, respectively. The stochastic term can be considered as a "perturbation by noise". So, clearly one motivation for studying SPDEs is to get information about the corresponding (unperturbed) deterministic PDE by letting the noise go to zero (e.g. replace B by $\varepsilon \cdot B$ and let $\varepsilon \to 0$) or to understand the different features occurring if one adds the noise term.

If we drop the stochastic term in these equations we get a deterministic PDE of "evolutionary type". Roughly speaking this means we have that the time derivative of the desired solution (on the left) is equal to a non–linear functional of its spatial derivatives (on the right).

Among others (see Subsection 4.1, in particular the cases, where Δ is replaced by the p-Laplacian) the approach presented in these notes will cover Examples 1.0.2 in case $d = 3$ or 4. (cf. Remark 4.1.10,2. and also [RRW06] without restrictions on the dimension) and 1.0.5 (cf. Example 4.1.11). For Example 1.0.1 we refer to [AR91] and for Examples 1.0.3 and 1.0.4 e.g. to [DPZ92], [DPZ96].

2. The Stochastic Integral in General Hilbert Spaces (w.r.t. Brownian Motion)

This chapter is a slight modification of Chap. 1 in [FK01].

We fix two separable Hilbert spaces $(U, \langle \, , \, \rangle_U)$ and $(H, \langle \, , \, \rangle)$. The first part of this chapter is devoted to the construction of the stochastic Itô integral

$$\int_0^t \Phi(s) \, \mathrm{d}W(s), \quad t \in [0, T],$$

where $W(t)$, $t \in [0, T]$, is a Wiener process on U and Φ is a process with values that are linear but not necessarily bounded operators from U to H.

For that we first will have to introduce the notion of the standard Wiener process in infinite dimensions. Then there will be a short section about martingales in general Hilbert spaces. These two concepts are important for the construction of the stochastic integral which will be explained in the following section.

In the second part of this chapter we will present the Itô formula and the stochastic Fubini theorem and establish basic properties of the stochastic integral, including the Burkholder–Davis–Gundy inequality.

Finally, we will describe how to transmit the definition of the stochastic integral to the case that $W(t)$, $t \in [0, T]$, is a cylindrical Wiener process. For simplicity we assume that U and H are real Hilbert spaces.

2.1. Infinite-dimensional Wiener processes

For a topological space X we denote its Borel σ-algebra by $\mathcal{B}(X)$.

Definition 2.1.1. A probability measure μ on $(U, \mathcal{B}(U))$ is called *Gaussian* if for all $v \in U$ the bounded linear mapping

$$v' : U \to \mathbb{R}$$

defined by

$$u \mapsto \langle u, v \rangle_U, \quad u \in U,$$

has a Gaussian law, i.e. for all $v \in U$ there exist $m := m(v) \in \mathbb{R}$ and $\sigma := \sigma(v) \in [0, \infty[$ such that, if $\sigma(v) > 0$,

$$\left(\mu \circ (v')^{-1}\right)(A) = \mu(v' \in A) = \frac{1}{\sqrt{2\pi\sigma^2}} \int_A e^{-\frac{(x-m)^2}{2\sigma^2}} \, \mathrm{d}x \quad \text{for all } A \in \mathcal{B}(\mathbb{R}),$$

and, if $\sigma(v) = 0$,

$$\mu \circ (v')^{-1} = \delta_{m(v)}.$$

Theorem 2.1.2. *A measure μ on $(U, \mathcal{B}(U))$ is Gaussian if and only if*

$$\hat{\mu}(u) := \int_U e^{i\langle u, v \rangle_U} \, \mu(\mathrm{d}v) = e^{i\langle m, u \rangle_U - \frac{1}{2}\langle Qu, u \rangle_U}, \quad u \in U,$$

where $m \in U$ and $Q \in L(U)$ is nonnegative, symmetric, with finite trace (see Definition B.0.3; here $L(U)$ denotes the set of all bounded linear operators on U).

In this case μ will be denoted by $N(m, Q)$ where m is called mean and Q is called covariance (operator). The measure μ is uniquely determined by m and Q.

Furthermore, for all $h, g \in U$

$$\int \langle x, h \rangle_U \, \mu(\mathrm{d}x) = \langle m, h \rangle_U,$$

$$\int \left(\langle x, h \rangle_U - \langle m, h \rangle_U\right)\left(\langle x, g \rangle_U - \langle m, g \rangle_U\right) \mu(\mathrm{d}x) = \langle Qh, g \rangle_U,$$

$$\int \|x - m\|_U^2 \, \mu(\mathrm{d}x) = \operatorname{tr} Q.$$

Proof. (cf. [DPZ92]) Obviously, a probability measure with this Fourier transform is Gaussian. Now let us conversely assume that μ is Gaussian. We need the following:

Lemma 2.1.3. *Let ν be a probability measure on $(U, \mathcal{B}(U))$. Let $k \in \mathbb{N}$ be such that*

$$\int_U |\langle z, x \rangle_U|^k \, \nu(\mathrm{d}x) < \infty \quad \forall z \in U.$$

Then there exists a constant $C = C(k, \nu) > 0$ such that for all $h_1, \ldots, h_k \in U$

$$\int_U |\langle h_1, x \rangle_U \cdots \langle h_k, x \rangle_U| \, \nu(\mathrm{d}x) \leqslant C \, \|h_1\|_U \cdots \|h_k\|_U.$$

In particular, the symmetric k-linear form

$$U^k \ni (h_1, \ldots, h_k) \mapsto \int \langle h_1, x \rangle_U \cdots \langle h_k, x \rangle_U \, \nu(\mathrm{d}x) \in \mathbb{R}$$

is continuous.

Proof. For $n \in \mathbb{N}$ define

$$U_n := \left\{ z \in U \;\middle|\; \int_U |\langle z, x \rangle_U|^k \, \nu(\mathrm{d}x) \leqslant n \right\}.$$

By assumption

$$U = \bigcup_{n=1}^{\infty} U_n.$$

Since U is a complete metric space, by the Baire category theorem, there exists $n_0 \in \mathbb{N}$ such that U_{n_0} has non-empty interior, so there exists a ball (with centre z_0 and radius r_0) $B(z_0, r_0) \subset U_{n_0}$. Hence

$$\int_U |\langle z_0 + y, x \rangle_U|^k \, \nu(\mathrm{d}x) \leqslant n_0 \quad \forall\, y \in B(0, r_0),$$

therefore for all $y \in B(0, r_0)$

$$\int_U |\langle y, x \rangle_U|^k \, \nu(\mathrm{d}x) = \int_U |\langle z_0 + y, x \rangle_U - \langle z_0, x \rangle_U|^k \, \nu(\mathrm{d}x)$$

$$\leqslant 2^{k-1} \int_U |\langle z_0 + y, x \rangle_U|^k \, \nu(\mathrm{d}x) + 2^{k-1} \int_U |\langle z_0, x \rangle_U|^k \, \nu(\mathrm{d}x)$$

$$\leqslant 2^k n_0.$$

Applying this for $y := r_0 z$, $z \in U$ with $|z|_U = 1$, we obtain

$$\int_U |\langle z, x \rangle_U|^k \, \nu(\mathrm{d}x) \leqslant 2^k n_0 r_0^{-k}.$$

Hence, if $h_1, \ldots, h_k \in U \setminus \{0\}$, then by the generalized Hölder inequality

$$\int_U \left| \left\langle \frac{h_1}{|h_1|_U}, x \right\rangle_U \cdots \left\langle \frac{h_k}{|h_k|_U}, x \right\rangle_U \right| \nu(\mathrm{d}x)$$

$$\leqslant \left(\int_U \left| \left\langle \frac{h_1}{|h_1|_U}, x \right\rangle_U \right|^k \nu(\mathrm{d}x) \right)^{1/k} \cdots \left(\int_U \left| \left\langle \frac{h_k}{|h_k|_U}, x \right\rangle_U \right|^k \nu(\mathrm{d}x) \right)^{1/k}$$

$$\leqslant 2^k n_0 r_0^{-k},$$

and the assertion follows. $\qquad\qquad\square$

Applying Lemma 2.1.3 for $k = 1$ and $\nu := \mu$ we obtain that

$$U \ni h \mapsto \int \langle h, x \rangle_U \, \mu(\mathrm{d}x) \in \mathbb{R}$$

is a continuous linear map, hence there exists $m \in U$ such that

$$\int_U \langle x, h \rangle_U \, \mu(\mathrm{d}x) = \langle m, h \rangle \quad \forall\, h \in H.$$

Applying Lemma 2.1.3 for $k = 2$ and $\nu := \mu$ we obtain that

$$U^2 \ni (h_1, h_2) \mapsto \int \langle x, h_1 \rangle_U \langle x, h_2 \rangle_U \, \mu(dx) - \langle m, h_1 \rangle_U \langle m, h_2 \rangle_U$$

is a continuous symmetric bilinear map, hence there exists a symmetric $Q \in L(U)$ such that this map is equal to

$$U^2 \ni (h_1, h_2) \mapsto \langle Qh_1, h_2 \rangle_U.$$

Since for all $h \in U$

$$\langle Qh, h \rangle_U = \int \langle x, h \rangle_U^2 \, \mu(dx) - \left(\int \langle x, h \rangle_U \, \mu(dx) \right)^2 \geqslant 0,$$

Q is positive definite. It remains to prove that Q is trace class (i.e.

$$\operatorname{tr} Q := \sum_{i=1}^{\infty} \langle Qe_i, e_i \rangle_U < \infty$$

for one (hence every) orthonormal basis $\{e_i \mid i \in \mathbb{N}\}$ of U, cf. Appendix B). We may assume without loss of generality that μ has mean zero, i.e. $m = 0$ ($\in U$), since the image measure of μ under the translation $U \ni x \mapsto x - m$ is again Gaussian with mean zero and the same covariance Q. Then we have for all $h \in U$ and all $c \in (0, \infty)$

$$1 - e^{-\frac{1}{2} \langle Qh, h \rangle_U} = \int_U \left(1 - \cos \langle h, x \rangle_U \right) \mu(dx)$$

$$\leqslant \int_{\{|x|_U \leqslant c\}} \left(1 - \cos \langle h, x \rangle_U \right) \mu(dx) + 2\mu(\{x \in U \mid |x|_U > c\})$$

$$\leqslant \frac{1}{2} \int_{\{|x|_U \leqslant c\}} \left| \langle h, x \rangle_U \right|^2 \mu(dx) + 2\mu(\{x \in U \mid |x|_U > c\}) \qquad (2.1.1)$$

(since $1 - \cos x \leqslant \frac{1}{2} x^2$). Defining the positive definite symmetric linear operator Q_c on U by

$$\langle Q_c h_1, h_2 \rangle_U := \int_{\{|x|_U \leqslant c\}} \langle h_1, x \rangle_U \cdot \langle h_2, x \rangle_U \, \mu(dx), \quad h_1, h_2 \in U,$$

we even have that Q_c is trace class because for every orthonormal basis $\{e_k \mid k \in \mathbb{N}\}$ of U we have (by monotone convergence)

$$\sum_{k=1}^{\infty} \langle Q_c e_k, e_k \rangle_U = \int_{\{|x|_U \leqslant c\}} \sum_{k=1}^{\infty} \langle e_k, x \rangle_U^2 \, \mu(dx) = \int_{\{|x|_U \leqslant c\}} |x|_U^2 \, \mu(dx)$$

$$\leqslant c^2 < \infty.$$

Claim: There exists $c_0 \in (0, \infty)$ (large enough) so that $Q \leqslant 2 \log 4 \, Q_{c_0}$ (meaning that $\langle Qh, h \rangle_U \leqslant 2 \log 4 \langle Q_{c_0} h, h \rangle_U$ for all $h \in U$).

To prove the claim let c_0 be so big that $\mu(\{x \in U \mid |x|_U > c_0\}) \leqslant \frac{1}{8}$. Let $h \in U$ such that $\langle Q_{c_0} h, h \rangle_U \leqslant 1$. Then (2.1.1) implies

$$1 - e^{-\frac{1}{2} \langle Qh, h \rangle_U} \leqslant \frac{1}{2} + \frac{1}{4} = \frac{3}{4},$$

hence $4 \geqslant e^{\frac{1}{2} \langle Qh, h \rangle_U}$, so $\langle Qh, h \rangle_U \leqslant 2 \log 4$. If $h \in U$ is arbitrary, but $\langle Q_{c_0} h, h \rangle_U \neq 0$, then we apply what we have just proved to $h / \langle Q_{c_0} h, h \rangle_U^{\frac{1}{2}}$ and the claim follows for such h. If, however, $\langle Q_{c_0} h, h \rangle = 0$, then for all $n \in \mathbb{N}$, $\langle Q_{c_0} nh, nh \rangle_U = 0 \leqslant 1$, hence by the above $\langle Qh, h \rangle_U \leqslant n^{-2} 2 \log 4$. Therefore, $\langle Q_{c_0} h, h \rangle_U = 0$ and the claim is proved, also for such h.

Since Q_{c_0} has finite trace, so has Q by the claim and the theorem is proved, since the uniqueness part follows from the fact that the Fourier transform is one-to-one. $\qquad\square$

The following result is then obvious.

Proposition 2.1.4. *Let X be a U-valued Gaussian random variable on a probability space $(\Omega, \mathcal{F}, P)$, i.e. there exist $m \in U$ and $Q \in L(U)$ nonnegative, symmetric, with finite trace such that $P \circ X^{-1} = N(m, Q)$.*

Then $\langle X, u \rangle_U$ is normally distributed for all $u \in U$ and the following statements hold:

- $E\big(\langle X, u \rangle_U\big) = \langle m, u \rangle_U$ *for all $u \in U$,*

- $E\big(\langle X - m, u \rangle_U \cdot \langle X - m, v \rangle_U\big) = \langle Qu, v \rangle_U$ *for all $u, v \in U$,*

- $E\big(\|X - m\|_U^2\big) = \operatorname{tr} Q.$

The following proposition will lead to a representation of a U-valued Gaussian random variable in terms of real-valued Gaussian random variables.

Proposition 2.1.5. *If $Q \in L(U)$ is nonnegative, symmetric, with finite trace then there exists an orthonormal basis e_k, $k \in \mathbb{N}$, of U such that*

$$Q e_k = \lambda_k e_k, \quad \lambda_k \geqslant 0, \ k \in \mathbb{N},$$

and 0 is the only accumulation point of the sequence $(\lambda_k)_{k \in \mathbb{N}}$.

Proof. See [RS72, Theorem VI.21; Theorem VI.16 (Hilbert–Schmidt theorem)]. $\qquad\square$

Proposition 2.1.6 (Representation of a Gaussian random variable). *Let $m \in U$ and $Q \in L(U)$ be nonnegative, symmetric, with $\operatorname{tr} Q < \infty$. In addition, we assume that e_k, $k \in \mathbb{N}$, is an orthonormal basis of U consisting of eigenvectors of Q with corresponding eigenvalues λ_k, $k \in \mathbb{N}$, as in Proposition 2.1.5, numbered in decreasing order.*

Then a U-valued random variable X on a probability space $(\Omega, \mathcal{F}, P)$ is Gaussian with $P \circ X^{-1} = N(m, Q)$ if and only if

$$X = \sum_{k \in \mathbb{N}} \sqrt{\lambda_k} \beta_k e_k + m \quad \text{(as objects in } L^2(\Omega, \mathcal{F}, P; U)),$$

where β_k, $k \in \mathbb{N}$, are independent real-valued random variables with $P \circ \beta_k^{-1} = N(0,1)$ for all $k \in \mathbb{N}$ with $\lambda_k > 0$. The series converges in $L^2(\Omega, \mathcal{F}, P; U)$.

Proof.

1. Let X be a Gaussian random variable with mean m and covariance Q. Below we set $\langle\,,\,\rangle := \langle\,,\,\rangle_U$.

   Then $X = \sum_{k \in \mathbb{N}} \langle X, e_k \rangle e_k$ in U where $\langle X, e_k \rangle$ is normally distributed with mean $\langle m, e_k \rangle$ and variance λ_k, $k \in \mathbb{N}$, by Proposition 2.1.4. If we now define

$$\beta_k : \begin{cases} = \dfrac{\langle X, e_k \rangle - \langle m, e_k \rangle}{\sqrt{\lambda_k}} & \text{if } k \in \mathbb{N} \text{ with } \lambda_k > 0 \\ \equiv 0 \in \mathbb{R} & \text{else,} \end{cases}$$

then we get that $P \circ \beta_k^{-1} = N(0,1)$ and $X = \sum_{k \in \mathbb{N}} \sqrt{\lambda_k} \beta_k e_k + m$. To prove the independence of β_k, $k \in \mathbb{N}$, we take an arbitrary $n \in \mathbb{N}$ and $a_k \in \mathbb{R}$, $1 \leqslant k \leqslant n$, and obtain that for $c := -\sum_{k=1, \lambda_k \neq 0}^{n} \frac{a_k}{\sqrt{\lambda_k}} \langle m, e_k \rangle$

$$\sum_{k=1}^{n} a_k \beta_k = \sum_{\substack{k=1, \\ \lambda_k \neq 0}}^{n} \frac{a_k}{\sqrt{\lambda_k}} \langle X, e_k \rangle + c = \left\langle X, \sum_{\substack{k=1, \\ \lambda_k \neq 0}}^{n} \frac{a_k}{\sqrt{\lambda_k}} e_k \right\rangle + c$$

which is normally distributed since X is a Gaussian random variable. Therefore we have that β_k, $k \in \mathbb{N}$, form a Gaussian family. Hence, to get the independence, we only have to check that the covariance of β_i and β_j, $i, j \in \mathbb{N}$, $i \neq j$, with $\lambda_i \neq 0 \neq \lambda_j$, is equal to zero. But this is clear since

$$E(\beta_i \beta_j) = \frac{1}{\sqrt{\lambda_i \lambda_j}} E(\langle X - m, e_i \rangle \langle X - m, e_j \rangle) = \frac{1}{\sqrt{\lambda_i \lambda_j}} \langle Q e_i, e_j \rangle$$

$$= \frac{\lambda_i}{\sqrt{\lambda_i \lambda_j}} \langle e_i, e_j \rangle = 0$$

for $i \neq j$.

Besides, the series $\sum_{k=1}^{n} \sqrt{\lambda_k} \beta_k e_k$, $n \in \mathbb{N}$, converges in $L^2(\Omega, \mathcal{F}, P; U)$ since the space is complete and

$$E\left(\left\| \sum_{k=m}^{n} \sqrt{\lambda_k} \beta_k e_k \right\|^2 \right) = \sum_{k=m}^{n} \lambda_k E(|\beta_k|^2) = \sum_{k=m}^{n} \lambda_k.$$

Since $\sum_{k \in \mathbb{N}} \lambda_k = \operatorname{tr} Q < \infty$ this expression becomes arbitrarily small for m and n large enough.

2. Let e_k, $k \in \mathbb{N}$, be an orthonormal basis of U such that $Qe_k = \lambda_k e_k$, $k \in \mathbb{N}$, and let β_k, $k \in \mathbb{N}$, be a family of independent real-valued Gaussian random variables with mean 0 and variance 1. Then it is clear that the series $\sum_{k=1}^{n} \sqrt{\lambda_k} \beta_k e_k + m$, $n \in \mathbb{N}$, converges to $X := \sum_{k \in \mathbb{N}} \sqrt{\lambda_k} \beta_k e_k + m$ in $L^2(\Omega, \mathcal{F}, P; U)$ (see part 1). Now we fix $u \in U$ and get that

$$\left\langle \sum_{k=1}^{n} \sqrt{\lambda_k} \beta_k e_k + m, u \right\rangle = \sum_{k=1}^{n} \sqrt{\lambda_k} \beta_k \langle e_k, u \rangle + \langle m, u \rangle$$

is normally distributed for all $n \in \mathbb{N}$ and the sequence converges in $L^2(\Omega, \mathcal{F}, P)$. This implies that the limit $\langle X, u \rangle$ is also normally distributed where

$$E(\langle X, u \rangle) = E\left(\sum_{k \in \mathbb{N}} \sqrt{\lambda_k} \beta_k \langle e_k, u \rangle + \langle m, u \rangle \right)$$

$$= \lim_{n \to \infty} E\left(\sum_{k=1}^{n} \sqrt{\lambda_k} \beta_k \langle e_k, u \rangle \right) + \langle m, u \rangle = \langle m, u \rangle$$

and concerning the covariance we obtain that

$$E\Big(\big(\langle X, u \rangle - \langle m, u \rangle \big) \big(\langle X, v \rangle - \langle m, v \rangle \big) \Big)$$

$$= \lim_{n \to \infty} E\left(\sum_{k=1}^{n} \sqrt{\lambda_k} \beta_k \langle e_k, u \rangle \sum_{k=1}^{n} \sqrt{\lambda_k} \beta_k \langle e_k, v \rangle \right)$$

$$= \sum_{k \in \mathbb{N}} \lambda_k \langle e_k, u \rangle \langle e_k, v \rangle = \sum_{k \in \mathbb{N}} \langle Qe_k, u \rangle \langle e_k, v \rangle$$

$$= \sum_{k \in \mathbb{N}} \langle e_k, Qu \rangle \langle e_k, v \rangle = \langle Qu, v \rangle$$

for all $u, v \in U$.

$\square$

By part 2 of this proof we finally get the following existence result.

Corollary 2.1.7. *Let Q be a nonnegative and symmetric operator in $L(U)$ with finite trace and let $m \in U$. Then there exists a Gaussian measure $\mu = N(m, Q)$ on $(U, \mathcal{B}(U))$.*

Let us give an alternative, more direct proof of Corollary 2.1.7 without using Proposition 2.1.6. For the proof we need the following exercise.

Exercise 2.1.8. *Consider $\mathbb{R}^\infty$ with the product topology. Let $\mathcal{B}(\mathbb{R}^\infty)$ denote its Borel σ-algebra. Prove:*

(i) $\mathcal{B}(\mathbb{R}^\infty) = \sigma(\pi_k \mid k \in \mathbb{N})$, where $\pi_k : \mathbb{R}^\infty \to \mathbb{R}$ denotes the projection on the k-th coordinate.

(ii) $l^2(\mathbb{R}) \ \left(:= \left\{ (x_k)_{k \in \mathbb{N}} \in \mathbb{R}^\infty \ \middle| \ \sum_{k=1}^\infty x_k^2 < \infty \right\} \right) \ \in \mathcal{B}(\mathbb{R}^\infty).$

(iii) $\mathcal{B}(\mathbb{R}^\infty) \cap l^2(\mathbb{R}) = \sigma(\pi_k|_{l^2} \mid k \in \mathbb{N}).$

(iv) Let $l^2(\mathbb{R})$ be equipped with its natural norm

$$\|x\|_{l^2} := \left(\sum_{k=1}^\infty x_k^2 \right)^{\frac{1}{2}}, \quad x = (x_k)_{k \in \mathbb{N}} \in l^2(\mathbb{R}),$$

and let $\mathcal{B}\big(l^2(\mathbb{R})\big)$ be the corresponding Borel σ-algebra. Then:

$$\mathcal{B}\big(l^2(\mathbb{R})\big) = \mathcal{B}(\mathbb{R}^\infty) \cap l^2(\mathbb{R}).$$

Alternative Proof of Corollary 2.1.7. It suffices to construct $N(0, Q)$, since $N(m, Q)$ is the image measure of $N(0, Q)$ under translation with m. For $k \in \mathbb{N}$ consider the normal distribution $N(0, \lambda_k)$ on $\mathbb{R}$ and let ν be their product measure on $\big(\mathbb{R}^\infty, \mathcal{B}(\mathbb{R}^\infty)\big)$, i.e.

$$\nu = \prod_{k \in \mathbb{N}} N(0, \lambda_k) \quad \text{on } \big(\mathbb{R}^\infty, \mathcal{B}(\mathbb{R}^\infty)\big).$$

Here λ_k, $k \in \mathbb{N}$, are as in Proposition 2.1.5. Since the map $g : \mathbb{R}^\infty \to [0, \infty]$ defined by

$$g(x) := \sum_{k=1}^\infty x_k^2, \quad x = (x_k)_{k \in \mathbb{N}} \in \mathbb{R}^\infty,$$

is $\mathcal{B}(\mathbb{R}^\infty)$-measurable, we may calculate

$$\int_{\mathbb{R}^\infty} g(x)\, \nu(\mathrm{d}x) = \sum_{k=1}^\infty \int x_k^2 \, N(0, \lambda_k)(\mathrm{d}x_k) = \sum_{k=1}^\infty \lambda_k < \infty.$$

Therefore, using Exercise 2.1.8(ii), we obtain $\nu\big(l^2(\mathbb{R})\big) = 1$. Restricting ν to $\mathcal{B}(\mathbb{R}^\infty) \cap l^2(\mathbb{R})$, by Exercise 2.1.8(iv) we get a probability measure, let us call it $\tilde{\mu}$, on $\big(l^2(\mathbb{R}), \mathcal{B}(l^2(\mathbb{R}))\big)$. Now take the orthonormal basis $\{e_k \mid k \in \mathbb{N}\}$ from Proposition 2.1.5 and consider the corresponding canonical isomorphism $I : l^2(\mathbb{R}) \to U$ defined by

$$I(x) = \sum_{k=1}^\infty x_k e_k, \quad x = (x_k)_{k \in \mathbb{N}} \in l^2(\mathbb{R}).$$

It is then easy to check that the image measure

$$\mu := \tilde{\mu} \circ I^{-1} \quad \text{on } \big(U, \mathcal{B}(U)\big)$$

is the desired measure, i.e. $\mu = N(0, Q)$. $\qquad\qquad\qquad\qquad\qquad\qquad \square$

After these preparations we will give the definition of the standard Q-Wiener process. To this end we fix an element $Q \in L(U)$, nonnegative, symmetric and with finite trace and a positive real number T.

Definition 2.1.9. A U-valued stochastic process $W(t)$, $t \in [0, T]$, on a probability space $(\Omega, \mathcal{F}, P)$ is called a (standard) Q-Wiener process if:

- $W(0) = 0$,

- W has P-a.s. continuous trajectories,

- the increments of W are independent, i.e. the random variables

$$W(t_1), \ W(t_2) - W(t_1), \ldots, \ W(t_n) - W(t_{n-1})$$

  are independent for all $0 \leqslant t_1 < \cdots < t_n \leqslant T$, $n \in \mathbb{N}$,

- the increments have the following Gaussian laws:

$$P \circ \big(W(t) - W(s)\big)^{-1} = N\big(0, (t-s)Q\big) \quad \text{for all } 0 \leqslant s \leqslant t \leqslant T.$$

Similarly to the existence of Gaussian measures the existence of a Q-Wiener process in U can be reduced to the real-valued case. This is the content of the following proposition.

Proposition 2.1.10 (Representation of the Q-Wiener process). *Let e_k, $k \in \mathbb{N}$, be an orthonormal basis of U consisting of eigenvectors of Q with corresponding eigenvalues λ_k, $k \in \mathbb{N}$. Then a U-valued stochastic process $W(t)$, $t \in [0, T]$, is a Q-Wiener process if and only if*

$$W(t) = \sum_{k \in \mathbb{N}} \sqrt{\lambda_k} \beta_k(t) e_k, \quad t \in [0, T], \tag{2.1.2}$$

where β_k, $k \in \{n \in \mathbb{N} \mid \lambda_n > 0\}$, are independent real-valued Brownian motions on a probability space $(\Omega, \mathcal{F}, P)$. The series even converges in $L^2\big(\Omega, \mathcal{F}, P; C([0, T], U)\big)$, and thus always has a P-a.s. continuous modification. (Here the space $C\big([0, T], U\big)$ is equipped with the sup norm.) In particular, for any Q as above there exists a Q-Wiener process on U.

Proof.

1. Let $W(t)$, $t \in [0, T]$, be a Q-Wiener process in U.

   Since $P \circ W(t)^{-1} = N(0, tQ)$, we see as in the proof of Proposition 2.1.6 that

   $$W(t) = \sum_{k \in \mathbb{N}} \sqrt{\lambda_k} \beta_k(t) e_k, \quad t \in [0, T],$$

   with

   $$\beta_k(t) : \begin{cases} = \frac{\langle W(t), e_k \rangle}{\sqrt{\lambda_k}} & \text{if } k \in \mathbb{N} \text{ with } \lambda_k > 0 \\ \equiv 0 & \text{else,} \end{cases}$$

for all $t \in [0, T]$. Furthermore, $P \circ \beta_k^{-1}(t) = N(0, t)$, $k \in \mathbb{N}$, and $\beta_k(t)$, $k \in \mathbb{N}$, are independent for each $t \in [0, T]$.

Now we fix $k \in \mathbb{N}$. First we show that $\beta_k(t)$, $t \in [0, T]$, is a Brownian motion:

If we take an arbitrary partition $0 = t_0 \leqslant t_1 < \cdots < t_n \leqslant T$, $n \in \mathbb{N}$, of $[0, T]$ we get that

$$\beta_k(t_1), \; \beta_k(t_2) - \beta_k(t_1), \ldots, \; \beta_k(t_n) - \beta_k(t_{n-1})$$

are independent for each $k \in \mathbb{N}$ since for $1 \leqslant j \leqslant n$

$$\beta_k(t_j) - \beta_k(t_{j-1}) = \begin{cases} \frac{1}{\sqrt{\lambda_k}}\langle W(t_j) - W(t_{j-1}), e_k \rangle & \text{if } \lambda_k > 0, \\ 0 & \text{else.} \end{cases}$$

Moreover, we obtain that for the same reason $P \circ \big(\beta_k(t) - \beta_k(s)\big)^{-1} = N(0, t - s)$ for $0 \leqslant s \leqslant t \leqslant T$.

In addition,

$$t \mapsto \frac{1}{\sqrt{\lambda_k}}\langle W(t), e_k \rangle = \beta_k(t)$$

is P-a.s. continuous for all $k \in \mathbb{N}$.

Secondly, it remains to prove that β_k, $k \in \mathbb{N}$, are independent.

We take $k_1, \ldots, k_n \in \mathbb{N}$, $n \in \mathbb{N}$, $k_i \neq k_j$ if $i \neq j$ and an arbitrary partition $0 = t_0 \leqslant t_1 \leqslant \ldots \leqslant t_m \leqslant T$, $m \in \mathbb{N}$.

Then we have to show that

$$\sigma\big(\beta_{k_1}(t_1), \ldots, \beta_{k_1}(t_m)\big), \ldots, \sigma\big(\beta_{k_n}(t_1), \ldots, \beta_{k_n}(t_m)\big)$$

are independent.

We will prove this by induction with respect to m:

If $m = 1$ it is clear that $\beta_{k_1}(t_1), \ldots, \beta_{k_n}(t_1)$ are independent as observed above. Thus, we now take a partition $0 = t_0 \leqslant t_1 \leqslant \ldots \leqslant t_{m+1} \leqslant T$ and assume that

$$\sigma\big(\beta_{k_1}(t_1), \ldots, \beta_{k_1}(t_m)\big), \ldots, \sigma\big(\beta_{k_n}(t_1), \ldots, \beta_{k_n}(t_m)\big)$$

are independent. We note that

$$\sigma\big(\beta_{k_i}(t_1), \ldots, \beta_{k_i}(t_m), \beta_{k_i}(t_{m+1})\big)$$
$$= \sigma\big(\beta_{k_i}(t_1), \ldots, \beta_{k_i}(t_m), \beta_{k_i}(t_{m+1}) - \beta_{k_i}(t_m)\big), \quad 1 \leqslant i \leqslant n,$$

and that

$$\beta_{k_i}(t_{m+1}) - \beta_{k_i}(t_m) = \begin{cases} \frac{1}{\sqrt{\lambda_{k_i}}}\langle W(t_{m+1}) - W(t_m), e_{k_i} \rangle_U & \text{if } \lambda_k > 0, \\ 0 & \text{else,} \end{cases}$$

$1 \leqslant i \leqslant n$, are independent since they are pairwise orthogonal in $L^2(\Omega, \mathcal{F}, P; \mathbb{R})$ and since $W(t_{m+1}) - W(t_m)$ is a Gaussian random variable. If we take $A_{i,j} \in \mathcal{B}(\mathbb{R})$, $1 \leqslant i \leqslant n$, $1 \leqslant j \leqslant m+1$, then because of the independence of $\sigma\big(W(s) \mid s \leqslant t_m\big)$ and $\sigma\big(W(t_{m+1}) - W(t_m)\big)$ we get that

$$P\Big(\bigcap_{i=1}^{n}\{\beta_{k_i}(t_1) \in A_{i,1}, \ldots, \beta_{k_i}(t_m) \in A_{i,m},$$

$$\beta_{k_i}(t_{m+1}) - \beta_{k_i}(t_m) \in A_{i,m+1}\}\Big)$$

$$=P\Big(\underbrace{\bigcap_{i=1}^{n}\bigcap_{j=1}^{m}\{\beta_{k_i}(t_j) \in A_{i,j}\}}_{\in\, \sigma(W(s)\,|\,s\,\leqslant\, t_m)} \cap \underbrace{\bigcap_{i=1}^{n}\{\beta_{k_i}(t_{m+1}) - \beta_{k_i}(t_m) \in A_{i,m+1}\}}_{\in\, \sigma(W(t_{m+1})\,-\,W(t_m))}\Big)$$

$$=P\Big(\bigcap_{i=1}^{n}\bigcap_{j=1}^{m}\{\beta_{k_i}(t_j) \in A_{i,j}\}\Big) \cdot P\Big(\bigcap_{i=1}^{n}\{\beta_{k_i}(t_{m+1}) - \beta_{k_i}(t_m) \in A_{i,m+1}\}\Big)$$

$$=\Big(\prod_{i=1}^{n} P\Big(\bigcap_{j=1}^{m}\{\beta_{k_i}(t_j) \in A_{i,j}\}\Big)\Big)$$

$$\cdot \Big(\prod_{i=1}^{n} P\{\beta_{k_i}(t_{m+1}) - \beta_{k_i}(t_m) \in A_{i,m+1}\}\Big)$$

$$=\prod_{i=1}^{n} P\Big(\bigcap_{j=1}^{m}\{\beta_{k_i}(t_j) \in A_{i,j}\} \cap \{\beta_{k_i}(t_{m+1}) - \beta_{k_i}(t_m) \in A_{i,m+1}\}\Big)$$

and therefore the assertion follows.

2. If we define
$$W(t) := \sum_{k \in \mathbb{N}} \sqrt{\lambda_k}\beta_k(t)e_k, \quad t \in [0,T],$$

where β_k, $k \in \mathbb{N}$, are independent real-valued continuous Brownian motions then it is clear that $W(t)$, $t \in [0,T]$, is well-defined in $L^2(\Omega, \mathcal{F}, P; U)$. Besides, it is obvious that the process $W(t)$, $t \in [0,T]$, starts at zero and that

$$P \circ \big(W(t) - W(s)\big)^{-1} = N\big(0, (t-s)Q\big), \quad 0 \leqslant s < t \leqslant T,$$

by Proposition 2.1.6. It is also clear that the increments are independent.

Thus it remains to show that the above series converges in $L^2\big(\Omega, \mathcal{F}, P; C([0,T], U)\big)$. To this end we set

$$W^N(t, \omega) := \sum_{k=1}^{N} \sqrt{\lambda_k}\beta_k(t, \omega)e_k$$

for all $(t, \omega) \in \Omega_T := [0, T] \times \Omega$ and $N \in \mathbb{N}$. Then W^N, $N \in \mathbb{N}$, is P-a.s. continuous and we have that for $M < N$

$$E\left(\sup_{t \in [0,T]} \left\| W^N(t) - W^M(t) \right\|_U^2 \right) = E\left(\sup_{t \in [0,T]} \sum_{k=M+1}^{N} \lambda_k \beta_k^2(t) \right)$$

$$\leqslant \sum_{k=M+1}^{N} \lambda_k E\left(\sup_{t \in [0,T]} \beta_k^2(t) \right) \leqslant c \sum_{k=M+1}^{N} \lambda_k$$

where $c_i = E\left(\sup_{t \in [0,T]} \beta_1^2(t) \right) < \infty$ because of Doob's maximal inequality for real-valued submartingales. As $\sum_{k \in \mathbb{N}} \lambda_k = \operatorname{tr} Q < \infty$, the assertion follows. $\qquad \square$

Definition 2.1.11 (Normal filtration). A filtration $\mathcal{F}_t$, $t \in [0, T]$, on a probability space $(\Omega, \mathcal{F}, P)$ is called normal if:

- $\mathcal{F}_0$ contains all elements $A \in \mathcal{F}$ with $P(A) = 0$ and

- $\mathcal{F}_t = \mathcal{F}_{t+} = \bigcap_{s > t} \mathcal{F}_s$ for all $t \in [0, T]$.

Definition 2.1.12 (Q-Wiener process with respect to a filtration).
A Q-Wiener process $W(t)$, $t \in [0, T]$, is called a Q-Wiener process with respect to a filtration $\mathcal{F}_t$, $t \in [0, T]$, if:

- $W(t)$, $t \in [0, T]$, is adapted to $\mathcal{F}_t$, $t \in [0, T]$, and

- $W(t) - W(s)$ is independent of $\mathcal{F}_s$ for all $0 \leqslant s \leqslant t \leqslant T$.

In fact it is possible to show that any U-valued Q-Wiener process $W(t)$, $t \in [0, T]$, is a Q-Wiener process with respect to a normal filtration:
We define

$$\mathcal{N} := \left\{ A \in \mathcal{F} \mid P(A) = 0 \right\}, \quad \tilde{\mathcal{F}}_t := \sigma\left(W(s) \mid s \leqslant t \right)$$

and $\quad \tilde{\mathcal{F}}_t^0 := \sigma(\tilde{\mathcal{F}}_t \cup \mathcal{N}).$

Then it is clear that

$$\mathcal{F}_t := \bigcap_{s > t} \tilde{\mathcal{F}}_s^0, \quad t \in [0, T], \tag{2.1.3}$$

is a normal filtration and we get:

Proposition 2.1.13. *Let $W(t)$, $t \in [0, T]$, be an arbitrary U-valued Q-Wiener process on a probability space $(\Omega, \mathcal{F}, P)$. Then it is a Q-Wiener process with respect to the normal filtration $\mathcal{F}_t$, $t \in [0, T]$, given by (2.1.3).*

Proof. It is clear that $W(t)$, $t \in [0, T]$, is adapted to $\mathcal{F}_t$, $t \in [0, T]$. Hence we only have to verify that $W(t) - W(s)$ is independent of $\mathcal{F}_s$, $0 \leqslant s < t \leqslant T$. But if we fix $0 \leqslant s < t \leqslant T$ it is clear that $W(t) - W(s)$ is independent of $\tilde{\mathcal{F}}_s$ since

$$\sigma\big(W(t_1), W(t_2), \ldots, W(t_n)\big)$$
$$= \sigma\big(W(t_1), W(t_2) - W(t_1), \ldots, W(t_n) - W(t_{n-1})\big)$$

for all $0 \leqslant t_1 < t_2 < \cdots < t_n \leqslant s$. Of course, $W(t) - W(s)$ is then also independent of $\tilde{\mathcal{F}}_s^0$. To prove now that $W(t) - W(s)$ is independent of $\mathcal{F}_s$ it is enough to show that

$$P\big(\{W(t) - W(s) \in A\} \cap B\big) = P\big(W(t) - W(s) \in A\big) \cdot P(B)$$

for any $B \in \mathcal{F}_s$ and any closed subset $A \subset U$ as $\mathcal{E} := \{A \subset U \mid A \text{ closed}\}$ generates $\mathcal{B}(U)$ and is stable under finite intersections. But we have

$$P\big(\{W(t) - W(s) \in A\} \cap B\big)$$
$$= E\Big(1_A \circ \big(W(t) - W(s)\big) \cdot 1_B\Big)$$
$$= \lim_{n \to \infty} E\Big(\big[\big(1 - n \operatorname{dist}(W(t) - W(s), A)\big) \vee 0\big] 1_B\Big)$$
$$= \lim_{n \to \infty} \lim_{m \to \infty} E\Big(\big[\big(1 - n \operatorname{dist}(W(t) - W(s + \tfrac{1}{m}), A)\big) \vee 0\big] 1_B\Big)$$
$$= \lim_{n \to \infty} \lim_{m \to \infty} E\Big(\big(1 - n \operatorname{dist}(W(t) - W(s + \tfrac{1}{m}), A)\big) \vee 0\Big) \cdot P(B)$$
$$= P\big(W(t) - W(s) \in A\big) \cdot P(B),$$

since $W(t) - W(s + \tfrac{1}{m})$ is independent of $\tilde{\mathcal{F}}_{s + \frac{1}{m}}^0 \supset \mathcal{F}_s$ if m is large enough. $\quad\square$

2.2. Martingales in general Banach spaces

Analogously to the real-valued case it is possible to define the conditional expectation of any Bochner integrable random variable with values in an arbitrary separable Banach space $(E, \| \, \|)$. This result is formulated in the following proposition.

Proposition 2.2.1 (Existence of the conditional expectation). *Assume that E is a separable real Banach space. Let X be a Bochner integrable E-valued random variable defined on a probability space $(\Omega, \mathcal{F}, P)$ and let $\mathcal{G}$ be a σ-field contained in $\mathcal{F}$.*

Then there exists a unique, up to a set of P-probability zero, Bochner integrable E-valued random variable Z, measurable with respect to $\mathcal{G}$ such that

$$\int_A X \, dP = \int_A Z \, dP \quad \text{for all } A \in \mathcal{G}. \tag{2.2.1}$$

The random variable Z is denoted by $E(X \mid \mathcal{G})$ and is called the conditional expectation *of X given $\mathcal{G}$. Furthermore,*

$$\left\| E(X \mid \mathcal{G}) \right\| \leqslant E(\|X\| \mid \mathcal{G}).$$

Proof. (cf. [DPZ92, Proposition 1.10, p. 27]) Let us first show uniqueness.

Since E is a separable Banach space, there exist $l_n \in E^*$, $n \in \mathbb{N}$, separating the points of E. Suppose that Z_1, Z_2 are Bochner integrable, $\mathcal{G}$-measurable mappings from Ω to E such that

$$\int_A X \, dP = \int_A Z_1 \, dP = \int_A Z_2 \, dP \quad \text{for all } A \in \mathcal{G}.$$

Then for $n \in \mathbb{N}$ by Proposition A.2.2

$$\int_A \left(l_n(Z_1) - l_n(Z_2) \right) \, dP = 0 \quad \text{for all } A \in \mathcal{G}.$$

Applying this with $A := \{ l_n(Z_1) > l_n(Z_2) \}$ and $A := \{ l_n(Z_1) < l_n(Z_2) \}$ it follows that $l_n(Z_1) = l_n(Z_2)$ P-a.s., so

$$\Omega_0 := \bigcap_{n \in \mathbb{N}} \{ l_n(Z_1) = l_n(Z_2) \}$$

has P-measure one. Since l_n, $n \in \mathbb{N}$, separate the points of E; it follows that $Z_1 = Z_2$ on Ω_0.

To show existence we first assume that X is a simple function. So, there exist $x_1, \ldots, x_N \in E$ and pairwise disjoint sets $A_1, \ldots, A_N \in \mathcal{F}$ such that

$$X = \sum_{k=1}^{N} x_k 1_{A_k}.$$

Define

$$Z := \sum_{k=1}^{N} x_k E(1_{A_k} \mid \mathcal{G}).$$

Then obviously Z is $\mathcal{G}$-measurable and satisfies (2.2.1). Furthermore,

$$\|Z\| \leqslant \sum_{k=1}^{N} \|x_k\| E(1_{A_k} \mid \mathcal{G}) = E\left(\sum_{k=1}^{N} \|x_k\| 1_{A_k} \,\middle|\, \mathcal{G} \right) = E(\|X\| \mid \mathcal{G}). \tag{2.2.2}$$

Taking expectation we get

$$E\big(\|Z\|\big) \leqslant E\big(\|X\|\big). \tag{2.2.3}$$

For general X take simple functions X_n, $n \in \mathbb{N}$, as in Lemma A.1.4 and define Z_n as above with X_n replacing X. Then by (2.2.3) for all $n, m \in \mathbb{N}$

$$E\big(\|Z_n - Z_m\|\big) \leqslant E\big(\|X_n - X_m\|\big),$$

so $Z := \lim_{n\to\infty} Z_n$ exists in $L^1(\Omega, \mathcal{F}, P; E)$. Therefore, for all $A \in \mathcal{G}$

$$\int_A X \, dP = \lim_{n\to\infty} \int_A X_n \, dP = \lim_{n\to\infty} \int_A Z_n \, dP = \int_A Z \, dP.$$

Clearly, Z can be chosen $\mathcal{G}$-measurable, since so are the Z_n. Furthermore, by (2.2.2)

$$\big\|E(X \mid \mathcal{G})\big\| = \|Z\| = \lim_{n\to\infty} \|Z_n\| \leqslant \lim_{n\to\infty} E\big(\|X_n\| \mid \mathcal{G}\big) = E\big(\|X\| \mid \mathcal{G}\big),$$

where the limits are taken in $L^1(P)$. $\qquad\square$

Later we will need the following result:

Proposition 2.2.2. *Let $(E_1, \mathcal{E}_1)$ and $(E_2, \mathcal{E}_2)$ be two measurable spaces and $\Psi : E_1 \times E_2 \to \mathbb{R}$ a bounded measurable function. Let X_1 and X_2 be two random variables on $(\Omega, \mathcal{F}, P)$ with values in $(E_1, \mathcal{E}_1)$ and $(E_2, \mathcal{E}_2)$ respectively, and let $\mathcal{G} \subset \mathcal{F}$ be a fixed σ-field.*

Assume that X_1 is $\mathcal{G}$-measurable and X_2 is independent of $\mathcal{G}$, then

$$E\big(\Psi(X_1, X_2) \mid \mathcal{G}\big) = \hat{\Psi}(X_1)$$

where

$$\hat{\Psi}(x_1) = E\big(\Psi(x_1, X_2)\big), \quad x_1 \in E_1.$$

Proof. A simple exercise or see [DPZ92, Proposition 1.12, p. 29]. $\qquad\square$

Remark 2.2.3. *The previous proposition can be easily extended to the case where the function Ψ is not necessarily bounded but nonnegative.*

Definition 2.2.4. Let $M(t)$, $t \geqslant 0$, be a stochastic process on $(\Omega, \mathcal{F}, P)$ with values in a separable Banach space E, and let $\mathcal{F}_t$, $t \geqslant 0$, be a filtration on $(\Omega, \mathcal{F}, P)$.

The process M is called an $\mathcal{F}_t$-*martingale*, if:

- $E\big(\|M(t)\|\big) < \infty$ for all $t \geqslant 0$,

- $M(t)$ is $\mathcal{F}_t$-measurable for all $t \geqslant 0$,

- $E\big(M(t) \mid \mathcal{F}_s\big) = M(s)$ P-a.s. for all $0 \leqslant s \leqslant t < \infty$.

Remark 2.2.5. *Let M be as above such that $E(\|M(t)\|) < \infty$ for all $t \in [0, T]$. Then M is an $\mathcal{F}_t$-martingale if and only if $l(M)$ is an $\mathcal{F}_t$-martingale for all $l \in E^*$. In particular, results like optional stopping etc. extend to E-valued martingales.*

There is the following connection to real-valued submartingales.

Proposition 2.2.6. *If $M(t)$, $t \geq 0$, is an E-valued $\mathcal{F}_t$-martingale and $p \in [1, \infty)$, then $\|M(t)\|^p$, $t \geq 0$, is a real-valued $\mathcal{F}_t$-submartingale.*

Proof. Since E is separable there exist $l_k \in E^*$, $k \in \mathbb{N}$, such that $\|z\| = \sup l_k(z)$ for all $z \in E$. Then for $s < t$

$$E(\|M_t\| \mid \mathcal{F}_s) \geq \sup_k E(l_k(M_t) \mid \mathcal{F}_s)$$

$$= \sup_k l_k(E(M_t \mid \mathcal{F}_s))$$

$$= \sup_k l_k(M_s) = \|M_s\|.$$

This proves the assertion for $p = 1$. Then Jensen's inequality implies the assertion for all $p \in [1, \infty)$. $\qquad\square$

Theorem 2.2.7 (Maximal inequality). *Let $p > 1$ and let E be a separable Banach space.*
If $M(t)$, $t \in [0, T]$, is a right-continuous E-valued $\mathcal{F}_t$-martingale, then

$$\left(E\left(\sup_{t \in [0,T]} \|M(t)\|^p \right) \right)^{\frac{1}{p}} \leq \frac{p}{p-1} \sup_{t \in [0,T]} \left(E(\|M(t)\|^p) \right)^{\frac{1}{p}}$$

$$= \frac{p}{p-1} \left(E(\|M(T)\|^p) \right)^{\frac{1}{p}}.$$

Proof. The inequality is a consequence of the previous proposition and Doob's maximal inequality for real-valued submartingales. $\qquad\square$

Remark 2.2.8. *We note that in the inequality in Theorem 2.2.7 the first norm is the standard norm on $L^p(\Omega, \mathcal{F}, P; C([0, T]; E))$, whereas the second is the standard norm on $C([0, T]; L^p(\Omega, \mathcal{F}, P; E))$. So, for right-continuous E-valued $\mathcal{F}_t$-martingales these two norms are equivalent.*

Now we fix $0 < T < \infty$ and denote by $\mathcal{M}_T^2(E)$ the space of all E-valued continuous, square integrable martingales $M(t)$, $t \in [0, T]$. This space will play an important role with regard to the definition of the stochastic integral. We will use especially the following fact.

Proposition 2.2.9. *The space $\mathcal{M}_T^2(E)$ equipped with the norm*

$$\|M\|_{\mathcal{M}_T^2} := \sup_{t \in [0,T]} \left(E\big(\|M(t)\|^2\big) \right)^{\frac{1}{2}} = \left(E\big(\|M(T)\|^2\big) \right)^{\frac{1}{2}}$$

$$\leqslant \left(E\big(\sup_{t \in [0,T]} \|M(t)\|^2 \big) \right)^{\frac{1}{2}} \leqslant 2 \cdot E\big(\|M(T)\|^2\big)^{\frac{1}{2}}.$$

is a Banach space.

Proof. By the Riesz–Fischer theorem the space $L^2\big(\Omega, \mathcal{F}, P; C([0,T], E)\big)$ is complete. So, we only have to show that $\mathcal{M}_T^2$ is closed. But this is obvious since even $L^1(\Omega, \mathcal{F}, P; E)$-limits of martingales are martingales. $\qquad \square$

Proposition 2.2.10. *Let $T > 0$ and $W(t)$, $t \in [0,T]$, be a U-valued Q-Wiener process with respect to a normal filtration $\mathcal{F}_t$, $t \in [0,T]$, on a probability space $(\Omega, \mathcal{F}, P)$. Then $W(t)$, $t \in [0,T]$, is a continuous square integrable $\mathcal{F}_t$-martingale, i.e. $W \in \mathcal{M}_T^2(U)$.*

Proof. The continuity is clear by definition and for each $t \in [0,T]$ we have that $E\big(\|W(t)\|_U^2\big) = t \operatorname{tr} Q < \infty$ (see Proposition 2.1.4). Hence let $0 \leqslant s \leqslant t \leqslant T$ and $A \in \mathcal{F}_s$. Then we get by Proposition A.2.2 that

$$\left\langle \int_A W(t) - W(s)\, \mathrm{d}P, u \right\rangle_U = \int_A \big\langle W(t) - W(s), u \big\rangle_U \, \mathrm{d}P$$

$$= P(A) \int \big\langle W(t) - W(s), u \big\rangle_U \, \mathrm{d}P = 0$$

for all $u \in U$ as $\mathcal{F}_s$ is independent of $W(t) - W(s)$ and $E\big(\langle W(t) - W(s), u \rangle_U\big) = 0$ for all $u \in U$. Therefore,

$$\int_A W(t)\, \mathrm{d}P = \int_A W(s) + \big(W(t) - W(s)\big)\, \mathrm{d}P$$

$$= \int_A W(s)\, \mathrm{d}P + \int_A W(t) - W(s)\, \mathrm{d}P$$

$$= \int_A W(s)\, \mathrm{d}P, \qquad \text{for all } A \in \mathcal{F}_s.$$

$$\square$$

2.3. The definition of the stochastic integral

For the whole section we fix a positive real number T and a probability space $(\Omega, \mathcal{F}, P)$ and we define $\Omega_T := [0,T] \times \Omega$ and $P_T := \mathrm{d}x \otimes P$ where $\mathrm{d}x$ is the Lebesgue measure.

Moreover, let $Q \in L(U)$ be symmetric, nonnegative and with finite trace and we consider a Q-Wiener process $W(t)$, $t \in [0,T]$, with respect to a normal filtration $\mathcal{F}_t$, $t \in [0,T]$.

2.3.1. Scheme of the construction of the stochastic integral

Step 1:   First we consider a certain class $\mathcal{E}$ of elementary $L(U, H)$-valued processes and define the mapping

$$\text{Int}: \quad \mathcal{E} \quad \rightarrow \quad \mathcal{M}_T^2(H) =: \mathcal{M}_T^2$$
$$\Phi \quad \mapsto \quad \int_0^t \Phi(s)\, dW(s), \quad t \in [0, T].$$

Step 2:   We prove that there is a certain norm on $\mathcal{E}$ such that

$$\text{Int} : \mathcal{E} \rightarrow \mathcal{M}_T^2$$

is an isometry. Since $\mathcal{M}_T^2$ is a Banach space this implies that Int can be extended to the abstract completion $\bar{\mathcal{E}}$ of $\mathcal{E}$. This extension remains isometric and it is unique.

Step 3:   We give an explicit representation of $\bar{\mathcal{E}}$.

Step 4:   We show how the definition of the stochastic integral can be extended by localization.

2.3.2. The construction of the stochastic integral in detail

Step 1:   First we define the class $\mathcal{E}$ of all elementary processes as follows.

Definition 2.3.1 (Elementary process). An $L = L(U, H)$-valued process $\Phi(t)$, $t \in [0, T]$, on $(\Omega, \mathcal{F}, P)$ with normal filtration $\mathcal{F}_t$, $t \in [0, T]$, is said to be *elementary* if there exist $0 = t_0 < \cdots < t_k = T$, $k \in \mathbb{N}$, such that

$$\Phi(t) = \sum_{m=0}^{k-1} \Phi_m \mathbb{1}_{]t_m, t_{m+1}]}(t), \quad t \in [0, T],$$

where:

- $\Phi_m : \Omega \rightarrow L(U, H)$ is $\mathcal{F}_{t_m}$-measurable, w.r.t. strong Borel σ-algebra on $L(U, H)$, $0 \leqslant m \leqslant k - 1$,

- Φ_m takes only a finite number of values in $L(U, H)$, $1 \leqslant m \leqslant k - 1$.

If we define now

$$\text{Int}(\Phi)(t) := \int_0^t \Phi(s)\, dW(s) := \sum_{m=0}^{k-1} \Phi_m \big(W(t_{m+1} \wedge t) - W(t_m \wedge t) \big), \quad t \in [0, T],$$

(this is obviously independent of the representation) for all $\Phi \in \mathcal{E}$, we have the following important result.

Proposition 2.3.2. *Let* $\Phi \in \mathcal{E}$*. Then the stochastic integral* $\int_0^t \Phi(s)\,\mathrm{d}W(s)$*,*
$t \in [0, T]$*, defined in the previous way, is a continuous square integrable mar-*
tingale with respect to $\mathcal{F}_t$*,* $t \in [0, T]$*, i.e.*

$$\mathrm{Int} : \mathcal{E} \to \mathcal{M}_T^2.$$

Proof. Let $\Phi \in \mathcal{E}$ be given by

$$\Phi(t) = \sum_{m=0}^{k-1} \Phi_m 1_{]t_m, t_{m+1}]}(t), \quad t \in [0, T],$$

as in Definition 2.3.1. Then it is clear that

$$t \mapsto \int_0^t \Phi(s)\,\mathrm{d}W(s) = \sum_{m=0}^{k-1} \Phi_m\big(W(t_{m+1} \wedge t) - W(t_m \wedge t)\big)$$

is P-a.s. continuous because of the continuity of the Wiener process and the
continuity of $\Phi_m(\omega) : U \to H$, $0 \leqslant m \leqslant k - 1$, $\omega \in \Omega$. In addition, we get for
each summand that

$$\Big\| \Phi_m\big(W(t_{m+1} \wedge t) - W(t_m \wedge t)\big) \Big\|$$

$$\leqslant \|\Phi_m\|_{L(U,H)} \big\| W(t_{m+1} \wedge t) - W(t_m \wedge t) \big\|_U.$$

Since $W(t)$, $t \in [0, T]$, is square integrable this implies that $\int_0^t \Phi(s)\,\mathrm{d}W(s)$ is
square integrable for each $t \in [0, T]$.

To prove the martingale property we take $0 \leqslant s \leqslant t \leqslant T$ and a set A from
$\mathcal{F}_s$. If $\{\Phi_m(\omega) \mid \omega \in \Omega\} := \{L_1^m, \ldots, L_{k_m}^m\}$ we obtain by Proposition A.2.2
and the martingale property of the Wiener process (more precisely using

optional stopping) that

$$\int_A \sum_{m=0}^{k-1} \Phi_m \big(W(t_{m+1} \wedge t) - W(t_m \wedge t) \big) \, \mathrm{d}P$$

$$= \sum_{\substack{0 \leqslant m \leqslant k-1, \\ t_{m+1} < s}} \int_A \Phi_m \big(W(t_{m+1} \wedge s) - W(t_m \wedge s) \big) \, \mathrm{d}P$$

$$+ \sum_{\substack{0 \leqslant m \leqslant k-1, \\ s \leqslant t_{m+1}}} \sum_{j=1}^{k_m} \int_{A \cap \{\Phi_m = L_j^m\}} L_j^m \big(W(t_{m+1} \wedge t) - W(t_m \wedge t) \big) \, \mathrm{d}P$$

$$= \sum_{\substack{0 \leqslant m \leqslant k-1, \\ t_{m+1} < s}} \int_A \Phi_m \big(W(t_{m+1} \wedge s) - W(t_m \wedge s) \big) \, \mathrm{d}P$$

$$+ \sum_{\substack{0 \leqslant m \leqslant k-1, \\ s \leqslant t_{m+1}}} \sum_{j=1}^{k_m} L_j^m \underbrace{\int_{A \cap \{\Phi_m = L_j^m\}}}_{\in \mathcal{F}_{s \vee t_m}} W(t_{m+1} \wedge t) - W(t_m \wedge t) \, \mathrm{d}P$$

$$= \sum_{\substack{0 \leqslant m \leqslant k-1, \\ t_{m+1} < s}} \int_A \Phi_m \big(W(t_{m+1} \wedge s) - W(t_m \wedge s) \big) \, \mathrm{d}P$$

$$+ \sum_{\substack{0 \leqslant m \leqslant k-1, \\ t_m < s \leqslant t_{m+1}}} \sum_{j=1}^{k_m} L_j^m \int_{A \cap \{\Phi_m = L_j^m\}} W(t_{m+1} \wedge s) - W(t_m \wedge s) \, \mathrm{d}P$$

$$= \int_A \sum_{m=0}^{k-1} \Phi_m \big(W(t_{m+1} \wedge s) - W(t_m \wedge s) \big) \, \mathrm{d}P. \qquad \square$$

Step 2: To verify the assertion that there is a norm on $\mathcal{E}$ such that Int : $\mathcal{E} \to \mathcal{M}_T^2$ is an isometry, we have to introduce the following notion.

Definition 2.3.3 (Hilbert–Schmidt operator). Let e_k, $k \in \mathbb{N}$, be an orthonormal basis of U. An operator $A \in L(U, H)$ is called Hilbert-Schmidt if

$$\sum_{k \in \mathbb{N}} \langle Ae_k, Ae_k \rangle < \infty.$$

In Appendix B we take a close look at this notion. So here we only summarize the results which are important for the construction of the stochastic integral.

The definition of a Hilbert–Schmidt operator and the number

$$\|A\|_{L_2} := \left(\sum_{k \in \mathbb{N}} \|Ae_k\|^2 \right)^{\frac{1}{2}}$$

are independent of the choice of the basis (see Remark B.0.6(i)). Moreover, the space $L_2(U, H)$ of all Hilbert–Schmidt operators from U to H equipped with the inner product

$$\langle A, B \rangle_{L_2} := \sum_{k \in \mathbb{N}} \langle Ae_k, Be_k \rangle$$

is a separable Hilbert space (see Proposition B.0.7). Later, we will use the fact that $\|A\|_{L^2(U,H)} = \|A^*\|_{L^2(H,U)}$, where A^* is the adjoint operator of A (see Remark B.0.6(i)). Furthermore, compositions of Hilbert–Schmidt with bounded linear operators are again Hilbert–Schmidt.

Besides we recall the following fact.

Proposition 2.3.4. *If $Q \in L(U)$ is nonnegative and symmetric then there exists exactly one element $Q^{\frac{1}{2}} \in L(U)$ nonnegative and symmetric such that $Q^{\frac{1}{2}} \circ Q^{\frac{1}{2}} = Q$.*

If, in addition, $\operatorname{tr} Q < \infty$ we have that $Q^{\frac{1}{2}} \in L_2(U)$ where $\|Q^{\frac{1}{2}}\|_{L_2}^2 = \operatorname{tr} Q$ and of course $L \circ Q^{\frac{1}{2}} \in L_2(U, H)$ for all $L \in L(U, H)$.
Proof. [RS72, Theorem VI.9, p. 196] □

After these preparations we simply calculate the $\mathcal{M}_T^2$-norm of

$$\int_0^t \Phi(s) \, dW(s), \, t \in [0, T],$$

and get the following result.

Proposition 2.3.5. *If $\Phi = \sum_{m=0}^{k-1} \Phi_m 1_{]t_m, t_{m+1}]}$ is an elementary $L(U, H)$-valued process then*

$$\left\| \int_0^{\cdot} \Phi(s) \, dW(s) \right\|_{\mathcal{M}_T^2}^2 = E \left(\int_0^T \|\Phi(s) \circ Q^{\frac{1}{2}}\|_{L_2}^2 \, ds \right) =: \|\Phi\|_T^2 \quad (\text{``Itô-isometry''}).$$

Proof. If we set $\Delta_m := W(t_{m+1}) - W(t_m)$ then we get that

$$\left\| \int_0^{\cdot} \Phi(s) \, dW(s) \right\|_{\mathcal{M}_T^2}^2 = E \left(\left\| \int_0^T \Phi(s) \, dW(s) \right\|_H^2 \right) = E \left(\left\| \sum_{m=0}^{k-1} \Phi_m \Delta_m \right\|_H^2 \right)$$

$$= E \left(\sum_{m=0}^{k-1} \|\Phi_m \Delta_m\|_H^2 \right) + 2E \left(\sum_{0 \leqslant m < n \leqslant k-1} \langle \Phi_m \Delta_m, \Phi_n \Delta_n \rangle_H \right).$$

Claim 1:

$$E \left(\sum_{m=0}^{k-1} \|\Phi_m \Delta_m\|_H^2 \right) = \sum_{m=0}^{k-1} (t_{m+1} - t_m) E \left(\|\Phi_m \circ Q^{\frac{1}{2}}\|_{L_2}^2 \right)$$

$$= \int_0^T E \left(\|\Phi(s) \circ Q^{\frac{1}{2}}\|_{L_2}^2 \right) ds.$$

To prove this we take an orthonormal basis f_k, $k \in \mathbb{N}$, of H and get by the Parseval identity and Levi's monotone convergence theorem that

$$E\big(\|\Phi_m \Delta_m\|_H^2\big) = \sum_{l \in \mathbb{N}} E\big(\langle \Phi_m \Delta_m, f_l \rangle_H^2\big) = \sum_{l \in \mathbb{N}} E\Big(E\big(\langle \Delta_m, \Phi_m^* f_l \rangle_U^2 \mid \mathcal{F}_{t_m}\big)\Big).$$

Taking an orthonormal basis e_k, $k \in \mathbb{N}$, of U we obtain that

$$\Phi_m^* f_l = \sum_{k \in \mathbb{N}} \langle f_l, \Phi_m e_k \rangle_H e_k.$$

Since $\langle f_l, \Phi_m e_k \rangle_H$ is $\mathcal{F}_{t_m}$-measurable, this implies that $\Phi_m^* f_l$ is $\mathcal{F}_{t_m}$-measurable by Proposition A.1.3. Using the fact that $\sigma(\Delta_m)$ is independent of $\mathcal{F}_{t_m}$ we obtain by Lemma 2.2.2 that for P-a.e. $\omega \in \Omega$

$$E\big(\langle \Delta_m, \Phi_m^* f_l \rangle_U^2 \mid \mathcal{F}_{t_m}\big)(\omega) = E\Big(\langle \Delta_m, \Phi_m^*(\omega) f_l \rangle_U^2\Big)$$

$$= (t_{m+1} - t_m)\Big\langle Q\big(\Phi_m^*(\omega) f_l\big), \Phi_m^*(\omega) f_l \Big\rangle_U,$$

since $E\big(\langle \Delta_m, u \rangle_U^2\big) = (t_{m+1} - t_m)\langle Qu, u \rangle_U$ for all $u \in U$. Thus, the symmetry of $Q^{\frac{1}{2}}$ finally implies that

$$E\big(\|\Phi_m \Delta_m\|_H^2\big) = \sum_{l \in \mathbb{N}} E\Big(E\big(\langle \Delta_m, \Phi_m^* f_l \rangle_U^2 \mid \mathcal{F}_{t_m}\big)\Big)$$

$$= (t_{m+1} - t_m) \sum_{l \in \mathbb{N}} E\big(\langle Q\Phi_m^* f_l, \Phi_m^* f_l \rangle_U\big)$$

$$= (t_{m+1} - t_m) \sum_{l \in \mathbb{N}} E\Big(\big\|Q^{\frac{1}{2}} \Phi_m^* f_l\big\|_U^2\Big)$$

$$= (t_{m+1} - t_m) E\Big(\big\|\big(\Phi_m \circ Q^{\frac{1}{2}}\big)^*\big\|_{L_2(H,U)}^2\Big)$$

$$= (t_{m+1} - t_m) E\Big(\big\|\Phi_m \circ Q^{\frac{1}{2}}\big\|_{L_2(U,H)}^2\Big).$$

Hence the first assertion is proved and it only remains to verify the following claim.

Claim 2:

$$E\big(\langle \Phi_m \Delta_m, \Phi_n \Delta_n \rangle_H\big) = 0, \quad 0 \leqslant m < n \leqslant k - 1.$$

But this can be proved in a similar way to Claim 1:

$$E\big(\langle \Phi_m \Delta_m, \Phi_n \Delta_n \rangle_H\big) = E\Big(E\big(\langle \Phi_n^* \Phi_m \Delta_m, \Delta_n \rangle_U \mid \mathcal{F}_{t_n}\big)\Big)$$

$$= \int E\Big(\langle \Phi_n^*(\omega) \Phi_m(\omega) \Delta_m(\omega), \Delta_n \rangle_U\Big) P(d\omega) = 0,$$

since $E(\langle u, \Delta_n \rangle_U) = 0$ for all $u \in U$ (see Proposition 2.2.2). Hence the assertion follows. $\qquad\square$

Hence the right norm on $\mathcal{E}$ has been identified. But strictly speaking $\| \ \|_T$ is only a seminorm on $\mathcal{E}$. Therefore, we have to consider equivalence classes of elementary processes with respect to $\| \ \|_T$ to get a norm on $\mathcal{E}$. For simplicity we will not change the notation but stress the following fact.

Remark 2.3.6. *If two elementary processes Φ and $\tilde{\Phi}$ belong to one equivalence class with respect to $\| \ \|_T$ it does not follow that they are equal P_T-a.e. because their values only have to correspond on $Q^{\frac{1}{2}}(U)$ P_T-a.e.*

Thus we finally have shown that

$$\text{Int} : \left(\mathcal{E}, \| \ \|_T\right) \to \left(\mathcal{M}_T^2, \| \ \|_{\mathcal{M}_T^2}\right)$$

is an isometric transformation. Since $\mathcal{E}$ is dense in the abstract completion $\bar{\mathcal{E}}$ of $\mathcal{E}$ with respect to $\| \ \|_T$ it is clear that there is a unique isometric extension of Int to $\bar{\mathcal{E}}$.

Step 3: To give an explicit representation of $\bar{\mathcal{E}}$ it is useful, at this moment, to introduce the subspace $U_0 := Q^{\frac{1}{2}}(U)$ with the inner product given by

$$\langle u_0, v_0 \rangle_0 := \left\langle Q^{-\frac{1}{2}} u_0, Q^{-\frac{1}{2}} v_0 \right\rangle_U,$$

$u_0, v_0 \in U_0$, where $Q^{-\frac{1}{2}}$ is the pseudo inverse of $Q^{\frac{1}{2}}$ in the case that Q is not one-to-one. Then we get by Proposition C.0.3(i) that $(U_0, \langle \ , \ \rangle_0)$ is again a separable Hilbert space.

The separable Hilbert space $L_2(U_0, H)$ is called L_2^0. By Proposition C.0.3(ii) we know that $Q^{\frac{1}{2}} g_k$, $k \in \mathbb{N}$, is an orthonormal basis of $(U_0, \langle \ , \ \rangle_0)$ if g_k, $k \in \mathbb{N}$, is an orthonormal basis of $(\text{Ker } Q^{\frac{1}{2}})^{\perp}$. This basis can be supplemented to a basis of U by elements of $\text{Ker } Q^{\frac{1}{2}}$. Thus we obtain that

$$\|L\|_{L_2^0} = \left\|L \circ Q^{\frac{1}{2}}\right\|_{L_2} \quad \text{for each } L \in L_2^0.$$

Define $L(U, H)_0 := \left\{ T|_{U_0} \mid T \in L(U, H) \right\}$. Since $Q^{\frac{1}{2}} \in L_2(U)$ it is clear that $L(U, H)_0 \subset L_2^0$ and that the $\| \ \|_T$-norm of $\Phi \in \mathcal{E}$ can be written in the following way:

$$\|\Phi\|_T = \left(E\left(\int_0^T \|\Phi(s)\|_{L_2^0}^2 \, ds \right) \right)^{\frac{1}{2}}$$

Besides we need the following σ-field:

$$\mathcal{P}_T := \sigma\Big(\big\{]s, t] \times F_s \mid 0 \leqslant s < t \leqslant T, \ F_s \in \mathcal{F}_s \big\} \cup \big\{ \{0\} \times F_0 \mid F_0 \in \mathcal{F}_0 \big\} \Big)$$

$$= \sigma\big(Y : \Omega_T \to \mathbb{R} \mid Y \text{ is left-continuous and adapted to}$$

$$\mathcal{F}_t, \ t \in [0, T] \big).$$

Let $\tilde{H}$ be an arbitrary separable Hilbert space. If $Y : \Omega_T \to \tilde{H}$ is $\mathcal{P}_T / \mathcal{B}(\tilde{H})$-measurable it is called ($\tilde{H}$-)predictable.

If, for example, the process Y itself is continuous and adapted to $\mathcal{F}_t$, $t \in [0, T]$, then it is predictable.

So, we are now able to characterize $\bar{\mathcal{E}}$.

Claim: There is an explicit representation of $\bar{\mathcal{E}}$ and it is given by

$$\mathcal{N}_W^2(0, T; H) := \left\{ \Phi : [0, T] \times \Omega \to L_2^0 \mid \Phi \text{ is predictable and } \|\Phi\|_T < \infty \right\}$$

$$= L^2 \left([0, T] \times \Omega, \mathcal{P}_T, dt \otimes P; L_2^0 \right).$$

For simplicity we also write $\mathcal{N}_W^2(0, T)$ or $\mathcal{N}_W^2$ instead of $\mathcal{N}_W^2(0, T; H)$.

To prove this claim we first notice the following facts:

1. Since $L(U, H)_0 \subset L_2^0$ and since any $\Phi \in \mathcal{E}$ is L_2^0-predictable by construction we have that $\mathcal{E} \subset \mathcal{N}_W^2$.

2. Because of the completeness of L_2^0 we get by Appendix A that

$$\mathcal{N}_W^2 = L^2(\Omega_T, \mathcal{P}_T, P_T; L_2^0)$$

is also complete.

Therefore $\mathcal{N}_W^2$ is at least a candidate for a representation of $\bar{\mathcal{E}}$. Thus there only remains to show that $\mathcal{E}$ is a dense subset of $\mathcal{N}_W^2$. But this is formulated in Proposition 2.3.8 below, which can be proved with the help of the following lemma.

Lemma 2.3.7. *There is an orthonormal basis of L_2^0 consisting of elements of $L(U, H)_0$. This implies especially that $L(U, H)_0$ is a dense subset of L_2^0.*

Proof. Since Q is symmetric, nonnegative and $\operatorname{tr} Q < \infty$ we know by Lemma 2.1.5 that there exists an orthonormal basis e_k, $k \in \mathbb{N}$, of U such that $Q e_k = \lambda_k e_k$, $\lambda_k \geqslant 0$, $k \in \mathbb{N}$. In this case $Q^{\frac{1}{2}} e_k = \sqrt{\lambda_k} e_k$, $k \in \mathbb{N}$ with $\lambda_k > 0$, is an orthonormal basis of U_0 (see Proposition C.0.3(ii)).

If f_k, $k \in \mathbb{N}$, is an orthonormal basis of H then by Proposition B.0.7 we know that

$$f_j \otimes \sqrt{\lambda_k} e_k = f_j \langle \sqrt{\lambda_k} e_k, \cdot \rangle_{U_0} = \frac{1}{\lambda_k} f_j \langle e_k, \cdot \rangle_U, \quad j, k \in \mathbb{N}, \lambda_k > 0,$$

form an orthonormal basis of L_0^2 consisting of operators in $L(U, H)$. But, of course,

$$\overline{\operatorname{span} \left(\frac{1}{\sqrt{\lambda_k}} f_j \otimes e_k \,\middle|\, j, k \in \mathbb{N} \text{ with } \lambda_k > 0 \right)} = L_2^0.$$

$\square$

Proposition 2.3.8. *If Φ is a L_2^0-predictable process such that $\|\Phi\|_T < \infty$ then there exists a sequence Φ_n, $n \in \mathbb{N}$, of $L(U,H)_0$-valued elementary processes such that*

$$\|\Phi - \Phi_n\|_T \longrightarrow 0 \quad as \ n \to \infty.$$

Proof. **Step 1:** If $\Phi \in \mathcal{N}_W^2$ there exists a sequence of simple random variables $\Phi_n = \sum_{k=1}^{M_n} L_k^n 1_{A_k^n}$, $A_k^n \in \mathcal{P}_T$ and $L_k^n \in L_2^0$, $n \in \mathbb{N}$, such that

$$\|\Phi - \Phi_n\|_T \longrightarrow 0 \quad as \ n \to \infty.$$

As L_2^0 is a Hilbert space this is a simple consequence of Lemma A.1.4 and Lebesgue's dominated convergence theorem.

Thus the assertion is reduced to the case that $\Phi = L 1_A$ where $L \in L_2^0$ and $A \in \mathcal{P}_T$.

Step 2: Let $A \in \mathcal{P}_T$ and $L \in L_2^0$. Then there exists a sequence L_n, $n \in \mathbb{N}$, in $L(U,H)_0$ such that

$$\|L 1_A - L_n 1_A\|_T \longrightarrow 0 \quad as \ n \to \infty.$$

This result is obvious by Lemma 2.3.7 and thus now we only have to consider the case that $\Phi = L 1_A$, $L \in L(U,H)_0$ and $A \in \mathcal{P}_T$.

Step 3: If $\Phi = L 1_A$, $L \in L(U,H)_0$, $A \in \mathcal{P}_T$, then there is a sequence Φ_n, $n \in \mathbb{N}$, of elementary $L(U,H)_0$-valued processes in the sense of Definition 2.3.1 such that

$$\|L 1_A - \Phi_n\|_T \longrightarrow 0 \quad as \ n \longrightarrow \infty.$$

To show this it is sufficient to prove that for any $\varepsilon > 0$ there is a finite union

$$\Lambda = \bigcup_{n=1}^{N} A_n \text{ of pairwise disjoint predictable rectangles}$$

$$A_n \in \left\{]s,t] \times F_s \ \middle| \ 0 \leqslant s < t \leqslant T, \ F_s \in \mathcal{F}_s \right\} \cup \left\{ \{0\} \times F_0 \ \middle| \ F_0 \in \mathcal{F}_0 \right\} =: \mathcal{A}$$

such that

$$P_T\big((A \setminus \Lambda) \cup (\Lambda \setminus A)\big) < \varepsilon.$$

For then we get that $\sum_{n=1}^{N} L 1_{A_n}$ differs from an elementary process by a function of type $1_{\{0\} \times F_0}$ with $F_0 \in \mathcal{F}_0$, which has $\|\cdot\|_T$-norm zero and

$$\left\| L 1_A - \sum_{n=1}^{N} L 1_{A_n} \right\|_T^2 = E\left(\int_0^T \left\| L \left(1_A - \sum_{n=1}^{N} 1_{A_n} \right) \right\|_{L_2^0}^2 \ ds \right) \leqslant \varepsilon \|L\|_{L_2^0}^2.$$

Hence we define

$$\mathcal{K} := \left\{ \bigcup_{i \in I} A_i \ \middle| \ I \text{ is finite and } A_i \in \mathcal{A}, \ i \in I \right\}.$$

Then $\mathcal{K}$ is an algebra and any element in $\mathcal{K}$ can be written as a finite disjoint union of elements in $\mathcal{A}$. Now let $\mathcal{G}$ be the family of all $A \in \mathcal{P}_T$ which can be approximated by elements of $\mathcal{K}$ in the above sense. Then $\mathcal{G}$ is a Dynkin system and therefore $\mathcal{P}_T = \sigma(\mathcal{K}) = \mathcal{D}(\mathcal{K}) \subset \mathcal{G}$ as $\mathcal{K} \subset \mathcal{G}$. $\qquad\square$

Step 4: Finally the so-called localization procedure provides the possibility to extend the definition of the stochastic integral even to the linear space

$$\mathcal{N}_W(0,T;H) := \left\{ \Phi : \Omega_T \to L_2^0 \,\middle|\, \Phi \text{ is predictable with} \right.$$

$$\left. P\left(\int_0^T \|\Phi(s)\|_{L_2^0}^2 \, ds < \infty \right) = 1 \right\}.$$

For simplicity we also write $\mathcal{N}_W(0,T)$ or $\mathcal{N}_W$ instead of $\mathcal{N}_W(0,T;H)$ and $\mathcal{N}_W$ is called the class of *stochastically integrable* processes on $[0,T]$.

The extension is done in the following way:

For $\Phi \in \mathcal{N}_W$ we define

$$\tau_n := \inf\left\{ t \in [0,T] \,\middle|\, \int_0^t \|\Phi(s)\|_{L_2^0}^2 \, ds > n \right\} \wedge T. \qquad (2.3.1)$$

Then by the right-continuity of the filtration $\mathcal{F}_t$, $t \in [0,T]$, we get that

$$\{\tau_n \leqslant t\} = \bigcap_{m \in \mathbb{N}} \left\{ \tau_n < t + \frac{1}{m} \right\}$$

$$= \bigcap_{m \in \mathbb{N}} \underbrace{\bigcup_{q \in [0, t+\frac{1}{m}[\, \cap \mathbb{Q}} \underbrace{\left\{ \int_0^q \|\Phi(s)\|_{L_2^0}^2 \, ds > n \right\}}_{\in \mathcal{F}_q \text{ by the real Fubini theorem}}}_{\in \mathcal{F}_{t+\frac{1}{m}} \text{ and decreasing in } m} \in \mathcal{F}_t.$$

Therefore τ_n, $n \in \mathbb{N}$, is an increasing sequence of stopping times with respect to $\mathcal{F}_t$, $t \in [0,T]$, such that

$$E\left(\int_0^T \|1_{]0,\tau_n]}(s)\Phi(s)\|_{L_2^0}^2 \, ds \right) \leqslant n < \infty.$$

In addition, the processes $1_{]0,\tau_n]}\Phi$, $n \in \mathbb{N}$, are still L_2^0-predictable since $1_{]0,\tau_n]}$ is left-continuous and $(\mathcal{F}_t)$-adapted or since

$$]0,\tau_n] := \left\{ (s,\omega) \in \Omega_T \,\middle|\, 0 < s \leqslant \tau_n(\omega) \right\}$$

$$= \left(\{ (s,\omega) \in \Omega_T \mid \tau_n(\omega) < s \leqslant T \} \cup \{0\} \times \Omega \right)^c$$

$$= \left(\underbrace{\bigcup_{q \in \mathbb{Q}} (]q,T] \times \underbrace{\{\tau_n \leqslant q\}}_{\in \mathcal{F}_q})}_{\in \mathcal{P}_T} \cup \{0\} \times \Omega \right)^c \in \mathcal{P}_T.$$

Thus we get that the stochastic integrals

$$\int_0^t 1_{]0,\tau_n]}(s)\Phi(s) \, dW(s), \quad t \in [0,T],$$

are well-defined for all $n \in \mathbb{N}$. For arbitrary $t \in [0, T]$ we set

$$\int_0^t \Phi(s) \, dW(s) := \int_0^t 1_{]0, \tau_n]}(s) \Phi(s) \, dW(s), \qquad (2.3.2)$$

where n is an arbitrary natural number such that $\tau_n \geqslant t$. (Note that the sequence τ_n, $n \in \mathbb{N}$, even reaches T P-a.s., in the sense that for P-a.e. $\omega \in \Omega$ there exists $n(\omega) \in \mathbb{N}$ such that $\tau_n(\omega) = T$ for all $n \geqslant n(\omega)$.)

To show that this definition is consistent we have to prove that for arbitrary natural numbers $m < n$ and $t \in [0, T]$

$$\int_0^t 1_{]0, \tau_m]}(s) \Phi(s) \, dW(s) = \int_0^t 1_{]0, \tau_n]}(s) \Phi(s) \, dW(s) \quad P\text{-a.s.}$$

on $\{\tau_m \geqslant t\} \subset \{\tau_n \geqslant t\}$. This result follows from the following lemma, which implies that the process in (2.3.2) is a continuous H-valued local martingale.

Lemma 2.3.9. *Assume that $\Phi \in \mathcal{N}_W^2$ and that τ is an $\mathcal{F}_t$-stopping time such that $P(\tau \leqslant T) = 1$. Then there exists a P-null set $N \in \mathcal{F}$ independent of $t \in [0, T]$ such that*

$$\int_0^t 1_{]0, \tau]}(s) \Phi(s) \, dW(s) = \mathrm{Int}\left(1_{]0, \tau]}\Phi\right)(t) = \mathrm{Int}(\Phi)(\tau \wedge t)$$

$$= \int_0^{\tau \wedge t} \Phi(s) \, dW(s) \quad \text{on } N^c \text{ for all } t \in [0, T].$$

Proof. Since both integrals which appear in the equation are P-a.s. continuous we only have to prove that they are equal P-a.s. at any fixed time $t \in [0, T]$.
Step 1: We first consider the case that $\Phi \in \mathcal{E}$ and that τ is a simple stopping time which means that it takes only a finite number of values.

Let $0 = t_0 < t_1 < \cdots < t_k \leqslant T$, $k \in \mathbb{N}$, and

$$\Phi = \sum_{m=0}^{k-1} \Phi_m 1_{]t_m, t_{m+1}]}$$

where $\Phi_m : \Omega \to L(U, H)$ is $\mathcal{F}_{t_m}$-measurable and only takes a finite number of values for all $0 \leqslant m \leqslant k - 1$.

If τ is a simple stopping time there exists $n \in \mathbb{N}$ such that $\tau(\Omega) = \{a_0, \ldots, a_n\}$ and

$$\tau = \sum_{j=0}^n a_j 1_{A_j}$$

where $0 \leqslant a_j < a_{j+1} \leqslant T$ and $A_j = \{\tau = a_j\} \in \mathcal{F}_{a_j}$. In this way we get that

$1_{]\tau,T]}\Phi$ is an elementary process since

$$1_{]\tau,T]}(s)\Phi(s) = \sum_{m=0}^{k-1} \Phi_m 1_{]t_m,t_{m+1}]\cap]\tau,T]}(s)$$

$$= \sum_{m=0}^{k-1}\sum_{j=0}^{n} 1_{A_j}\Phi_m 1_{]t_m,t_{m+1}]\cap]a_j,T]}(s)$$

$$= \sum_{m=0}^{k-1}\sum_{j=0}^{n} \underbrace{1_{A_j}\Phi_m}_{\mathcal{F}_{t_m\vee a_j}\text{-measurable}} 1_{]t_m\vee a_j,t_{m+1}\vee a_j]}(s)$$

and concerning the integral we are interested in, we obtain that

$$\int_0^t 1_{]0,\tau]}(s)\Phi(s)\,dW(s) = \int_0^t \Phi(s)\,dW(s) - \int_0^t 1_{]\tau,T]}(s)\Phi(s)\,dW(s)$$

$$= \sum_{m=0}^{k-1} \Phi_m\big(W(t_{m+1}\wedge t) - W(t_m\wedge t)\big)$$

$$\quad - \sum_{m=0}^{k-1}\sum_{j=0}^{n} 1_{A_j}\Phi_m\Big(W\big((t_{m+1}\vee a_j)\wedge t\big) - W\big((t_m\vee a_j)\wedge t\big)\Big)$$

$$= \sum_{m=0}^{k-1} \Phi_m\big(W(t_{m+1}\wedge t) - W(t_m\wedge t)\big)$$

$$\quad - \sum_{m=0}^{k-1}\sum_{j=0}^{n} 1_{A_j}\Phi_m\Big(W\big((t_{m+1}\vee \tau)\wedge t\big) - W\big((t_m\vee \tau)\wedge t\big)\Big)$$

$$= \sum_{m=0}^{k-1} \Phi_m\big(W(t_{m+1}\wedge t) - W(t_m\wedge t)\big)$$

$$\quad - \sum_{m=0}^{k-1} \Phi_m\Big(W\big((t_{m+1}\vee \tau)\wedge t\big) - W\big((t_m\vee \tau)\wedge t\big)\Big)$$

$$= \sum_{m=0}^{k-1} \Phi_m\Big(W(t_{m+1}\wedge t) - W(t_m\wedge t)$$

$$\quad - W\big((t_{m+1}\vee \tau)\wedge t\big) - W\big((t_m\vee \tau)\wedge t\big)\Big)$$

$$= \sum_{m=0}^{k-1} \Phi_m\Big(W(t_{m+1}\wedge \tau\wedge t) - W(t_m\wedge \tau\wedge t)\Big) = \int_0^{t\wedge\tau} \Phi(s)\,dW(s).$$

Step 2: Now we consider the case that Φ is still an elementary process while τ is an arbitrary stopping time with $P(\tau \leqslant T) = 1$.

Then there exists a sequence

$$\tau_n = \sum_{k=0}^{2^n-1} T(k+1)2^{-n} 1_{]Tk2^{-n},T(k+1)2^{-n}]} \circ \tau, \quad n \in \mathbb{N},$$

of simple stopping times such that $\tau_n \downarrow \tau$ as $n \to \infty$ and because of the continuity of the stochastic integral we get that

$$\int_0^{\tau_n \wedge t} \Phi(s)\, dW(s) \xrightarrow{n \to \infty} \int_0^{\tau \wedge t} \Phi(s)\, dW(s) \quad P\text{-a.s.}$$

Besides, we obtain (even for non-elementary processes Φ) that

$$\left\| 1_{]0,\tau_n]}\Phi - 1_{]0,\tau]}\Phi \right\|_T^2 = E\left(\int_0^T 1_{]\tau,\tau_n]}(s)\|\Phi(s)\|_{L_2^0}^2 \, ds \right) \xrightarrow{n \to \infty} 0,$$

which by the definition of the integral implies that

$$E\left(\left\| \int_0^t 1_{]0,\tau_n]}(s)\Phi(s)\, dW(s) - \int_0^t 1_{]0,\tau]}(s)\Phi(s)\, dW(s) \right\|^2 \right) \xrightarrow{n \to \infty} 0$$

for all $t \in [0, T]$. As by Step 1

$$\int_0^t 1_{]0,\tau_n]}(s)\Phi(s)\, dW(s) = \int_0^{\tau_n \wedge t} \Phi(s)\, dW(s), \quad n \in \mathbb{N},\ t \in [0, T],$$

the assertion follows.

Step 3: Finally we generalize the statement to arbitrary $\Phi \in \mathcal{N}_W^2(0, T)$:

If $\Phi \in \mathcal{N}_W^2(0, T)$ then there exists a sequence of elementary processes Φ_n, $n \in \mathbb{N}$, such that

$$\|\Phi_n - \Phi\|_T \xrightarrow{n \to \infty} 0 \,.$$

By the definition of the stochastic integral this means that

$$\int_0^{\cdot} \Phi_n(s)\, dW(s) \xrightarrow{n \to \infty} \int_0^{\cdot} \Phi(s)\, dW(s) \quad \text{in } \mathcal{M}_T^2 \,.$$

Hence it follows that there is a subsequence n_k, $k \in \mathbb{N}$, and a P-null set $N \in \mathcal{F}$ independent of $t \in [0, T]$ such that

$$\int_0^t \Phi_{n_k}(s)\, dW(s) \xrightarrow{k \to \infty} \int_0^t \Phi(s)\, dW(s) \quad \text{on } N^c$$

for all $t \in [0, T]$ and therefore we get for all $t \in [0, T]$ that

$$\int_0^{\tau \wedge t} \Phi_{n_k}(s)\, dW(s) \xrightarrow{k \to \infty} \int_0^{\tau \wedge t} \Phi(s)\, dW(s) \quad P\text{-a.s.}$$

In addition, it is clear that

$$\| 1_{]0,\tau]} \Phi_n - 1_{]0,\tau]} \Phi \|_T \xrightarrow[n \to \infty]{} 0$$

which implies that for all $t \in [0, T]$

$$E \left(\left\| \int_0^t 1_{]0,\tau]}(s) \Phi_n(s) \, dW(s) - \int_0^t 1_{]0,\tau]}(s) \Phi(s) \, dW(s) \right\|^2 \right) \xrightarrow[n \to \infty]{} 0.$$

As by Step 2

$$\int_0^t 1_{]0,\tau]}(s) \Phi_{n_k}(s) \, dW(s) = \int_0^{\tau \wedge t} \Phi_{n_k}(s) \, dW(s) \quad P\text{-a.s.}$$

for all $k \in \mathbb{N}$ the assertion follows. $\square$

Therefore, for $m < n$ on $\{\tau_m \geqslant t\} \subset \{\tau_n \geqslant t\}$

$$\int_0^t 1_{]0,\tau_n]}(s) \Phi(s) \, dW(s) = \int_0^{\tau_m \wedge t} 1_{]0,\tau_n]}(s) \Phi(s) \, dW(s)$$

$$= \int_0^t 1_{]0,\tau_m]}(s) 1_{]0,\tau_n]}(s) \Phi(s) \, dW(s) = \int_0^t 1_{]0,\tau_m]}(s) \Phi(s) \, dW(s) \quad P\text{-a.s.,}$$

where we used Lemma 2.3.9 for the second equality. Hence the definition is consistent.

Remark 2.3.10. *In fact it is easy to see that the definition of the stochastic integral does not depend on the choice of τ_n, $n \in \mathbb{N}$. If σ_n, $n \in \mathbb{N}$, is another sequence of stopping times such that $\sigma_n \uparrow T$ as $n \to \infty$ and $1_{]0,\sigma_n]} \Phi \in \mathcal{N}_W^2$ for all $n \in \mathbb{N}$ we also get that*

$$\int_0^t \Phi(s) \, dW(s) = \lim_{n \to \infty} \int_0^t 1_{]0,\sigma_n]}(s) \Phi(s) \, dW(s) \quad P\text{-a.s. for all } t \in [0, T].$$

Proof. Let $t \in [0, T]$. Then we get that on the set $\{\tau_m \geqslant t\}$

$$\int_0^t \Phi(s) \, dW(s) = \int_0^t 1_{]0,\tau_m]}(s) \Phi(s) \, dW(s)$$

$$= \lim_{n \to \infty} \int_0^{t \wedge \sigma_n} 1_{]0,\tau_m]}(s) \Phi(s) \, dW(s)$$

$$= \lim_{n \to \infty} \int_0^{t \wedge \tau_m} 1_{]0,\sigma_n]}(s) \Phi(s) \, dW(s)$$

$$= \lim_{n \to \infty} \int_0^t 1_{]0,\sigma_n]}(s) \Phi(s) \, dW(s) \quad P\text{-a.s..} \quad \square$$

2.4. Properties of the stochastic integral

Let T be a positive real number and $W(t)$, $t \in [0, T]$, a Q-Wiener process as described at the beginning of the previous section.

Lemma 2.4.1. *Let Φ be a L_2^0-valued stochastically integrable process, $(\tilde{H}, \| \ \|_{\tilde{H}})$ a further separable Hilbert space and $L \in L(H, \tilde{H})$.*
Then the process $L(\Phi(t))$, $t \in [0, T]$, is an element of $\mathcal{N}_W(0, T; \tilde{H})$ and

$$L\left(\int_0^T \Phi(t) \, dW(t) \right) = \int_0^T L(\Phi(t)) \, dW(t) \quad P\text{-a.s.}$$

Proof. Since Φ is a stochastically integrable process and

$$\left\| L(\Phi(t)) \right\|_{L_2(U_0, \tilde{H})} \leqslant \|L\|_{L(H, \tilde{H})} \|\Phi(t)\|_{L_2^0},$$

it is obvious that $L(\Phi(t))$, $t \in [0, T]$, is $L_2(U_0, \tilde{H})$-predictable and

$$P\left(\int_0^T \left\| L(\Phi(t)) \right\|_{L_2(U_0, \tilde{H})}^2 \, dt < \infty \right) = 1.$$

Step 1: As the first step we consider the case that Φ is an elementary process, i.e.

$$\Phi(t) = \sum_{m=0}^{k-1} \Phi_m 1_{]t_m, t_{m+1}]}(t), \quad t \in [0, T],$$

where $0 = t_0 < t_1 < \cdots < t_k = T$, $\Phi_m : \Omega \to L(U, H)$ $\mathcal{F}_{t_m}$-measurable with $|\Phi_m(\Omega)| < \infty$ for $0 \leqslant m \leqslant k$. Then

$$L\left(\int_0^T \Phi(t) \, dW(t) \right) = L\left(\sum_{m=0}^{k-1} \Phi_m \big(W(t_{m+1}) - W(t_m) \big) \right)$$

$$= \sum_{m=0}^{k-1} L\big(\Phi_m \big(W(t_{m+1}) - W(t_m) \big) \big) = \int_0^T L(\Phi(t)) \, dW(t).$$

Step 2: Now let $\Phi \in \mathcal{N}_W^2(0, T)$. Then there exists a sequence Φ_n, $n \in \mathbb{N}$, of elementary processes with values in $L(U, H)_0$ such that

$$\|\Phi_n - \Phi\|_T = \left(E\left(\int_0^T \|\Phi_n(t) - \Phi(t)\|_{L_2^0}^2 \, dt \right) \right)^{\frac{1}{2}} \xrightarrow{n \to \infty} 0.$$

Then $L(\Phi_n)$, $n \in \mathbb{N}$, is a sequence of elementary processes with values in $L(U, \tilde{H})_0$ and

$$\big\| L(\Phi_n) - L(\Phi) \big\|_T \leqslant \|L\|_{L(H, \tilde{H})} \|\Phi_n - \Phi\|_T \xrightarrow{n \to \infty} 0.$$

By the definition of the stochastic integral, Step 1 and the continuity of L we get that there is a subsequence n_k, $k \in \mathbb{N}$, such that

$$\int_0^T L\big(\Phi(t)\big) \, dW(t) = \lim_{k \to \infty} \int_0^T L\big(\Phi_{n_k}(t)\big) \, dW(t)$$

$$= \lim_{k \to \infty} L\left(\int_0^T \Phi_{n_k}(t) \, dW(t)\right) = L\left(\lim_{k \to \infty} \int_0^T \Phi_{n_k}(t) \, dW(t)\right)$$

$$= L\left(\int_0^T \Phi(t) \, dW(t)\right) \quad P\text{-a.s.}$$

Step 3:   Finally let $\Phi \in \mathcal{N}_W(0, T)$.

Let τ_n, $n \in \mathbb{N}$, be a sequence of stopping times such that $\tau_n \uparrow T$ as $n \to \infty$ and $1_{]0,\tau_n]} \Phi \in \mathcal{N}_W^2(0, T, H)$. Then $1_{]0,\tau_n]} L(\Phi) \in \mathcal{N}_W^2(0, T, \tilde{H})$ for all $n \in \mathbb{N}$ and we obtain by Remark 2.3.10 and Step 2 (selecting a subsequence if necessary)

$$\int_0^T L\big(\Phi(t)\big) \, dW(t) = \lim_{n \to \infty} \int_0^T 1_{]0,\tau_n]}(t) L\big(\Phi(t)\big) \, dW(t)$$

$$= \lim_{n \to \infty} L\left(\int_0^T 1_{]0,\tau_n]}(t) \Phi(t) \, dW(t)\right) = L\left(\lim_{n \to \infty} \int_0^T 1_{]0,\tau_n]}(t) \Phi(t) \, dW(t)\right)$$

$$= L\left(\int_0^T \Phi(t) \, dW(t)\right) \quad P\text{-a.s.}$$

$\square$

Lemma 2.4.2. *Let $\Phi \in \mathcal{N}_W(0, T)$ and f an $(\mathcal{F}_t)$-adapted continuous H-valued process. Set*

$$\int_0^T \langle f(t), \Phi(t) \, dW(t) \rangle := \int_0^T \tilde{\Phi}_f(t) \, dW(t) \tag{2.4.1}$$

with

$$\tilde{\Phi}_f(t)(u) := \langle f(t), \Phi(t) u \rangle, \quad u \in U_0.$$

Then the stochastic integral in (2.4.1) is well-defined as a continuous $\mathbb{R}$-valued stochastic process. More precisely, $\tilde{\Phi}_f$ is a $\mathcal{P}_T / \mathcal{B}(L_2(U_0, \mathbb{R}))$-measurable map from $[0, T] \times \Omega$ to $L_2(U_0, \mathbb{R})$,

$$\|\tilde{\Phi}_f(t, \omega)\|_{L_2(U_0, \mathbb{R})} = \|\Phi^*(t, \omega) f(t, \omega)\|_{U_0}$$

for all $(t, \omega) \in [0, T] \times \Omega$ and

$$\int_0^T \|\tilde{\Phi}_f(t)\|_{L_2(U_0, \mathbb{R})}^2 \, dt \leqslant \sup_{t \in [0,T]} \|f(t)\| \int_0^T \|\Phi(t)\|_{L_2^0}^2 \, dt < \infty \quad P\text{-a.e..}$$

Proof. Since f is continuous, $\tilde{\Phi}_f$ is clearly predictable. Let e_k, $k \in \mathbb{N}$, be an orthonormal basis of U_0. Then for all $(t, \omega) \in [0, T] \times \Omega$

$$
\begin{aligned}
\|\tilde{\Phi}_f(t, \omega)\|^2_{L_2(U_0, \mathbb{R})} &= \sum_{k=1}^{\infty} \langle f(t, \omega), \Phi(t, \omega) e_k \rangle^2 \\
&= \sum_{k=1}^{\infty} \langle \Phi^*(t, \omega) f(t, \omega), e_k \rangle^2_{U_0} \\
&= \|\Phi^*(t, \omega) f(t, \omega)\|^2_{U_0} \\
&\leqslant \|\Phi^*(t, \omega)\|^2_{L(H, U_0)} \|f(t, \omega)\|^2_H \\
&\leqslant \|\Phi^*(t, \omega)\|^2_{L_2(H, U_0)} \|f(t, \omega)\|^2_H \\
&= \|\Phi(t, \omega)\|^2_{L_2^0} \|f(t, \omega)\|^2_H,
\end{aligned}
$$

where we used Remark B.0.6(i) in the last step. Now all assertions follow. $\quad\square$

Lemma 2.4.3. *Let $\Phi \in \mathcal{N}_W(0, T)$ and $M(t) := \int_0^t \Phi(s) \, dW(s)$, $t \in [0, T]$. Define*

$$
\langle M \rangle_t := \int_0^t \|\Phi(s)\|^2_{L_2^0} \, ds, \; t \in [0, T].
$$

Then $\langle M \rangle$ is the unique continuous increasing $(\mathcal{F}_t)$-adapted process starting at zero such that $\|M(t)\|^2 - \langle M \rangle_t$, $t \in [0, T]$, is a local martingale. If $\Phi \in \mathcal{N}^2_W(0, T)$, then for any sequence

$$
I_l := \{0 = t_0^l < t_1^l < \ldots < t_{k_l}^l = T\}, \; l \in \mathbb{N},
$$

of partitions with

$$
\max_i (t_i^l - t_{i-1}^l) \to 0 \; \text{as } l \to \infty
$$

$$
\lim_{l \to \infty} E \left(\left| \sum_{t_{j+1}^l \leqslant t} \|M(t_{j+1}^l) - M(t_j^l)\|^2 - \langle M \rangle_t \right| \right) = 0.
$$

Proof. For $n \in \mathbb{N}$ let τ_n be as in (2.3.1) and τ an $\mathcal{F}_t$-stopping time with $P[\tau \leqslant T] = 1$. Then by Lemma 2.3.9 for $\sigma := \tau \wedge \tau_n$, $t \in [0, T]$

$$
\begin{aligned}
E \left(\left\| \int_0^{t \wedge \sigma} \Phi(s) \, dW(s) \right\|^2 \right) &= E \left(\left\| \int_0^t 1_{]0, \sigma]} \Phi(s) \, dW(s) \right\|^2 \right) \\
&= E \left(\int_0^t \|1_{]0, \sigma]} \Phi(s)\|^2_{L_2^0} \, ds \right) \\
&= E \left(\int_0^{t \wedge \sigma} \|\Phi(s)\|^2_{L_2^0} \, ds \right),
\end{aligned}
$$

and the first assertion follows, because the uniqueness is obvious, since any real-valued local martingale of bounded variation is constant.

To prove the second assertion we fix an orthonormal basis $\{e_i | i \in \mathbb{N}\}$ of H and note that by the theory of real-valued martingales we have for each $i \in \mathbb{N}$

$$\lim_{l \to \infty} E\left(\left| \sum_{t^l_{j+1} \leqslant t} \langle e_i, M(t^l_{j+1}) - M(t^l_j) \rangle^2_H - \int_0^t \|\Phi(s)^* e_i\|^2_{U_0} \, ds \right| \right) = 0, \quad (2.4.2)$$

since by the first part of the assertion and Lemmas 2.4.1 and 2.4.2

$$\left\langle \int_0^t \langle e_i, \Phi(s) \, dW(s) \rangle_H \right\rangle_t = \int_0^t \|\Phi(s)^* e_i\|^2_{U_0} \, ds, \ t \in [0, T].$$

Furthermore, for all $i \in \mathbb{N}$

$$E\left(\left| \sum_{t^l_{j+1} \leqslant t} \langle e_i, M(t^l_{j+1}) - M(t^l_j) \rangle^2_H - \int_0^t \|\Phi(s)^* e_i\|^2_{U_0} \, ds \right| \right)$$

$$\leqslant \sum_{t^l_{j+1} \leqslant t} E\left[\left(\int_{t^l_j}^{t^l_{j+1}} \langle e_i, \Phi(s) \, dW(s) \rangle_H \right)^2 \right] + E\left(\int_0^t \|\Phi(s)^* e_i\|^2_{U_0} \, ds \right)$$

$$= \sum_{t^l_{j+1} \leqslant t} E\left(\int_{t^l_j}^{t^l_{j+1}} \|\Phi(s)^* e_i\|^2_{U_0} \, ds \right) + E\left(\int_0^t \|\Phi(s)^* e_i\|^2_{U_0} \, ds \right)$$

$$\leqslant 2E\left(\int_0^t \|\Phi(s)^* e_i\|^2_{U_0} \, ds \right)$$

$$(2.4.3)$$

which is summable over $i \in \mathbb{N}$. Here we used the isometry property of Int in the second to last step. But

$$E\left(\left| \sum_{t^l_{j+1} \leqslant t} \|M(t^l_{j+1}) - M(t^l_j)\|^2 - \int_0^t \|\Phi(s)\|^2_{L_2^0} \, ds \right| \right)$$

$$= E\left(\left| \sum_{i=1}^{\infty} \left(\sum_{t^l_{j+1} \leqslant t} \langle e_i, M(t^l_{j+1}) - M(t^l_j) \rangle^2_H - \int_0^t \|\Phi(s)^* e_i\|^2_{U_0} \, ds \right) \right| \right)$$

$$\leqslant \sum_{i=1}^{\infty} E\left(\left| \sum_{t^l_{j+1} \leqslant t} \langle e_i, M(t^l_{j+1}) - M(t^l_j) \rangle^2_H - \int_0^t \|\Phi(s)^* e_i\|^2_{U_0} \, ds \right| \right)$$

where we used Remark B.0.6(i) in the second step. Hence the second assertion follows by Lebesgue dominated convergence theorem from (2.4.2) and (2.4.3).

$\square$

2.5. The stochastic integral for cylindrical Wiener processes

Until now we have considered the case that $W(t)$, $t \in [0,T]$, was a standard Q-Wiener process where $Q \in L(U)$ was nonnegative, symmetric and with finite trace. We could integrate processes in

$$\mathcal{N}_W := \left\{ \Phi : \Omega_T \to L_2(Q^{\frac{1}{2}}(U), H) \mid \Phi \text{ is predictable and} \right.$$

$$\left. P\left(\int_0^T \|\Phi(s)\|_{L_2^0}^2 \, ds < \infty \right) = 1 \right\}.$$

In fact it is possible to extend the definition of the stochastic integral to the case that Q is not necessarily of finite trace. To this end we first have to introduce the concept of cylindrical Wiener processes.

2.5.1. Cylindrical Wiener processes

Let $Q \in L(U)$ be nonnegative definite and symmetric. Remember that in the case that Q is of finite trace the Q-Wiener process has the following representation:

$$W(t) = \sum_{k \in \mathbb{N}} \beta_k(t) e_k, \quad t \in [0,T],$$

where e_k, $k \in \mathbb{N}$, is an orthonormal basis of $Q^{\frac{1}{2}}(U) = U_0$ and β_k, $k \in \mathbb{N}$, is a family of independent real-valued Brownian motions. The series converges in $L^2(\Omega, \mathcal{F}, P; U)$, because the inclusion $U_0 \subset U$ defines a Hilbert–Schmidt embedding from $(U_0, \langle \, , \, \rangle_0)$ to $(U, \langle \, , \, \rangle)$. In the case that Q is no longer of finite trace one looses this convergence. Nevertheless, it is possible to define the Wiener process.

To this end we need a further Hilbert space $(U_1, \langle \, , \, \rangle_1)$ and a Hilbert–Schmidt embedding

$$J : (U_0, \langle \, , \, \rangle_0) \to (U_1, \langle \, , \, \rangle_1).$$

Remark 2.5.1. $(U_1, \langle \, , \, \rangle_1))$ *and* J *as above always exist; e.g. choose* $U_1 := U$ *and* $\alpha_k \in]0, \infty[$, $k \in \mathbb{N}$, *such that* $\sum_{k=1}^{\infty} \alpha_k^2 < \infty$. *Define* $J : U_0 \to U$ *by*

$$J(u) := \sum_{k=1}^{\infty} \alpha_k \langle u, e_k \rangle_0 \, e_k, \quad u \in U_0.$$

Then J *is one-to-one and Hilbert–Schmidt.*

Then the process given by the following proposition is called a *cylindrical Q-Wiener process* in U.

Proposition 2.5.2. *Let e_k, $k \in \mathbb{N}$, be an orthonormal basis of $U_0 = Q^{\frac{1}{2}}(U)$ and β_k, $k \in \mathbb{N}$, a family of independent real-valued Brownian motions. Define $Q_1 := JJ^*$. Then $Q_1 \in L(U_1)$, Q_1 is nonnegative definite and symmetric with finite trace and the series*

$$W(t) = \sum_{k=1}^{\infty} \beta_k(t) Je_k, \quad t \in [0, T], \tag{2.5.1}$$

converges in $\mathcal{M}_T^2(U_1)$ and defines a Q_1-Wiener process on U_1. Moreover, we have that $Q_1^{\frac{1}{2}}(U_1) = J(U_0)$ and for all $u_0 \in U_0$

$$\|u_0\|_0 = \|Q_1^{-\frac{1}{2}} Ju_0\|_1 = \|Ju_0\|_{Q_1^{\frac{1}{2}} U_1},$$

i.e. $J : U_0 \to Q_1^{\frac{1}{2}} U_1$ is an isometry.

Proof. **Step 1:** We prove that $W(t)$, $t \in [0, T]$, defined in (2.5.1) is a Q_1-Wiener process in U_1.

If we set $\xi_j(t) := \beta_j(t) J(e_j)$, $j \in \mathbb{N}$, we obtain that $\xi_j(t)$, $t \in [0, T]$, is a continuous U_1-valued martingale with respect to

$$\mathcal{G}_t := \sigma\left(\bigcup_{j \in \mathbb{N}} \sigma(\beta_j(s) | s \leqslant t) \right),$$

$t \in [0, T]$, since

$$E(\beta_j(t) \mid \mathcal{G}_s) = E(\beta_j(t) \mid \sigma(\beta_j(u) | u \leqslant s)) = \beta_j(s) \quad \text{for all } 0 \leqslant s < t \leqslant T$$

as $\sigma\big(\sigma(\beta_j(u) | u \leqslant s) \cup \sigma(\beta_j(t))\big)$ is independent of

$$\sigma\left(\bigcup_{\substack{k \in \mathbb{N} \\ k \neq j}} \sigma(\beta_k(u) | u \leqslant s) \right).$$

Then it is clear that

$$W_n(t) := \sum_{j=1}^{n} \beta_j(t) J(e_j), \quad t \in [0, T],$$

is also a continuous U_1-valued martingale with respect to $\mathcal{G}_t$, $t \in [0, T]$. In addition, we obtain that

$$E\left(\sup_{t \in [0,T]} \left\| \sum_{j=n}^{m} \beta_j(t) J(e_j) \right\|_1^2 \right) \leqslant 4 \sup_{t \in [0,T]} E\left(\left\| \sum_{j=n}^{m} \beta_j(t) J(e_j) \right\|_1^2 \right)$$

$$= 4T \sum_{j=n}^{m} \|J(e_j)\|_1^2, \quad m \geqslant n \geqslant 1.$$

Note that $\|J\|^2_{L_2(U_0,U_1)} = \sum_{j\in\mathbb{N}}\|J(e_j)\|^2_1 < \infty$. Therefore, we get the convergence of $W_n(t)$, $t \in [0,T]$, in $\mathcal{M}^2_T(U_1)$, hence the limit $W(t)$, $t \in [0,T]$, is P-a.s. continuous.

Now we want to show that $P \circ (W(t) - W(s))^{-1} = N(0,(t-s)JJ^*)$. Analogously to the second part of the proof of Proposition 2.1.6 we get that $\langle W(t) - W(s), u_1\rangle_1$ is normally distributed for all $0 \leqslant s < t \leqslant T$ and $u_1 \in U_1$. It is easy to see that the mean is equal to zero and concerning the covariance of $\langle W(t) - W(s), u_1\rangle_1$ and $\langle W(t) - W(s), v_1\rangle_1$, $u_1, v_1 \in U_1$, we obtain that

$$E(\langle W(t) - W(s), u_1\rangle_1\langle W(t) - W(s), v_1\rangle_1)$$

$$= \sum_{k\in\mathbb{N}}(t-s)\langle Je_k, u_1\rangle_1\langle Je_k, v_1\rangle_1$$

$$= (t-s)\sum_{k\in\mathbb{N}}\langle e_k, J^*u_1\rangle_0\langle e_k, J^*v_1\rangle_0$$

$$= (t-s)\langle J^*u_1, J^*v_1\rangle_0 = (t-s)\langle JJ^*u_1, v_1\rangle_1.$$

Thus, it only remains to show that the increments of $W(t)$, $t \in [0,T]$, are independent but this can be done in the same way as in the proof of Proposition 2.1.10.

Step 2: We prove that $\operatorname{Im} Q_1^{\frac{1}{2}} = J(U_0)$ and that $\|u_0\|_0 = \|Q_1^{-\frac{1}{2}}Ju_0\|_1$ for all $u_0 \in U_0$.

Since $Q_1 = JJ^*$, by Corollary C.0.6 we obtain that $Q_1^{\frac{1}{2}}(U_1) = J(U_0)$ and that $\|Q_1^{-\frac{1}{2}}u_1\|_1 = \|J^{-1}u_1\|_0$ for all $u_1 \in J(U_0)$. We now replace u_1 by $J(u_0)$, $u_0 \in U_0$, to get the last assertion, because $J : U_0 \to U_1$ is one-to-one. $\square$

2.5.2. The definition of the stochastic integral for cylindrical Wiener processes

We fix $Q \in L(U)$ nonnegative, symmetric but not necessarily of finite trace. After the preparations of the previous section we are now able to define the stochastic integral with respect to a cylindrical Q-Wiener process $W(t)$, $t \in [0,T]$.

Basically we integrate with respect to the standard U_1-valued Q_1-Wiener process given by Proposition 2.5.2. In this sense we first get that a process $\Phi(t)$, $t \in [0,T]$, is integrable with respect to $W(t)$, $t \in [0,T]$, if it takes values in $L_2(Q_1^{\frac{1}{2}}(U_1), H)$, is predictable and if

$$P\left(\int_0^T \|\Phi(s)\|^2_{L_2(Q_1^{\frac{1}{2}}(U_1),H)}\, ds < \infty\right) = 1.$$

But in addition, we have by Proposition 2.5.2 that $Q_1^{\frac{1}{2}}(U_1) = J(U_0)$ and that

$$\langle Ju_0, Jv_0 \rangle_{Q_1^{\frac{1}{2}}(U_1)} = \langle Q_1^{-\frac{1}{2}} Ju_0, Q_1^{-\frac{1}{2}} Jv_0 \rangle_1 = \langle u_0, v_0 \rangle_0$$

for all $u_0, v_0 \in U_0$ (by polarization). In particular, it follows that Je_k, $k \in \mathbb{N}$, is an orthonormal basis of $Q_1^{\frac{1}{2}}(U_1)$. Hence we get that

$$\Phi \in L_2^0 = L_2(Q^{\frac{1}{2}}(U), H) \Longleftrightarrow \Phi \circ J^{-1} \in L_2(Q_1^{\frac{1}{2}}(U_1), H)$$

since

$$\|\Phi\|_{L_2^0}^2 = \sum_{k \in \mathbb{N}} \langle \Phi e_k, \Phi e_k \rangle$$

$$= \sum_{k \in \mathbb{N}} \langle \Phi \circ J^{-1}(Je_k), \Phi \circ J^{-1}(Je_k) \rangle = \|\Phi \circ J^{-1}\|_{L_2(Q_1^{\frac{1}{2}}(U_1), H)}^2$$

Now we define

$$\int_0^t \Phi(s) \ dW(s) := \int_0^t \Phi(s) \circ J^{-1} \ dW(s), \quad t \in [0, T]. \tag{2.5.2}$$

Then the class of all integrable processes is given by

$$\mathcal{N}_W = \left\{ \Phi : \Omega_T \to L_2^0 \mid \Phi \text{ predictable and } P\left(\int_0^T \|\Phi(s)\|_{L_2^0}^2 \ ds < \infty \right) = 1 \right\}$$

as in the case where $W(t)$, $t \in [0, T]$, is a standard Q-Wiener process in U.

Remark 2.5.3.

1. *We note that the stochastic integral defined in (2.5.2) is independent of the choice of $(U_1, \langle \ , \ \rangle_1)$ and J. This follows by construction, since by (2.5.1) for elementary processes (2.5.2) does not depend on J.*

2. *If $Q \in L(U)$ is nonnegative, symmetric and with finite trace the standard Q-Wiener process can also be considered as a cylindrical Q-Wiener process by setting $J = I : U_0 \to U$ where I is the identity map. In this case both definitions of the stochastic integral coincide.*

Finally, we note that since the stochastic integrals in this chapter all have a standard Wiener process as integrator, we can drop the predictability assumption on $\Phi \in \mathcal{N}_W$ and just assume progressive measurability, i.e. $\Phi|_{[0,t] \times \Omega}$ is $\mathcal{B}([0, t]) \otimes \mathcal{F}_t / \mathcal{B}(L_2^0)$-measurable for all $t \in [0, T]$, at least if $(\Omega, \mathcal{F}, P)$ is complete (otherwise we consider its completion) (cf. [WW90, Theorem 6.3.1]).

We used the above framework so that it easily extends to more general Hilbert-space-valued martingales as integrators replacing the standard Wiener process. Details are left to the reader.

3. Stochastic Differential Equations in Finite Dimensions

This chapter is an extended version of [Kry99, Section 1].

3.1. Main result and a localization lemma

Let $(\Omega, \mathcal{F}, P)$ be a complete probability space and $\mathcal{F}_t$, $t \in [0, \infty[$, a normal filtration. Let $(W_t)_{t \geqslant 0}$ be a standard Wiener process on $\mathbb{R}^{d_1}$, $d_1 \in \mathbb{N}$, with respect to $\mathcal{F}_t$, $t \in [0, \infty[$. So, in the terminology of the previous section $U := \mathbb{R}^{d_1}$, $Q := \mathrm{I}$. The role of the Hilbert space H there will be taken by $\mathbb{R}^d$, $d \in \mathbb{N}$.

Let $M(d \times d_1, \mathbb{R})$ denote the set of all real $d \times d_1$-matrices. Let the following maps $\sigma = \sigma(t, x, \omega)$, $b = b(t, x, \omega)$ be given:

$$\sigma : [0, \infty[\times \mathbb{R}^d \times \Omega \to M(d \times d_1, \mathbb{R})$$

$$b : [0, \infty[\times \mathbb{R}^d \times \Omega \to \mathbb{R}^d$$

such that both are continuous in $x \in \mathbb{R}^d$ for each fixed $t \in [0, \infty[$, $w \in \Omega$, and progressively measurable, i.e. for each t their restriction to $[0, t] \times \Omega$ is $\mathcal{B}([0, t]) \otimes \mathcal{F}_t$-measurable, for each fixed $x \in \mathbb{R}^d$. We note that then both σ and b restricted to $[0, t] \times \mathbb{R}^d \times \Omega$ are $\mathcal{B}([0, t]) \otimes \mathcal{B}(\mathbb{R}^d) \otimes \mathcal{F}_t$-measurable for every $t \in [0, \infty[$. In particular, for every $x \in \mathbb{R}^d$, $t \in [0, \infty[$ both are $\mathcal{F}_t$-measurable. We also assume that the following integrability conditions hold:

$$\int_0^T \sup_{|x| \leqslant R} \{\|\sigma(t, x)\|^2 + |b(t, x)|\} \, dt < \infty \text{ on } \Omega, \tag{3.1.1}$$

for all $T, R \in [0, \infty[$. Here $|\cdot|$ denotes the Euclidean distance on $\mathbb{R}^d$ and

$$\|\sigma\|^2 := \sum_{i=1}^d \sum_{j=1}^{d_1} |\sigma_{ij}|^2. \tag{3.1.2}$$

$\langle \, , \, \rangle$ below denotes the Euclidean inner product on $\mathbb{R}^d$.

Theorem 3.1.1. *Let b, σ be as above satisfying (3.1.1). Assume that on Ω for all $t, R \in [0, \infty[$, $x, y \in \mathbb{R}^d$, $|x|, |y| \leqslant R$*

$$2\langle x - y, b(t, x) - b(t, y)\rangle + \|\sigma(t, x) - \sigma(t, y)\|^2$$

(local weak monotonicity)

$$\leqslant K_t(R)|x - y|^2$$

$$\text{(3.1.3)}$$

and

$$2\langle x, b(t, x)\rangle + \|\sigma(t, x)\|^2 \leqslant K_t(1)(1 + |x|^2), \quad \text{(weak coercivity)} \quad (3.1.4)$$

where for $R \in [0, \infty[$, $K_t(R)$ is an $\mathbb{R}_+$-valued $(\mathcal{F}_t)$-adapted process satisfying on Ω for all $R, T \in [0, \infty[$

$$\alpha_T(R) := \int_0^T K_t(R) \, \mathrm{d}t < \infty. \tag{3.1.5}$$

Then for any $\mathcal{F}_0$-measurable map $X_0 : \Omega \to \mathbb{R}^d$ there exists a (up to P-indistinguish-ability) unique solution to the stochastic differential equation

$$\mathrm{d}X(t) = b(t, X(t)) \, \mathrm{d}t + \sigma(t, X(t)) \, \mathrm{d}W(t). \tag{3.1.6}$$

Here solution means that $(X(t))_{t \geqslant 0}$ is a P-a.s. continuous $\mathbb{R}^d$-valued $(\mathcal{F}_t)$-adapted process such that P-a.s. for all $t \in [0, \infty[$

$$X(t) = X_0 + \int_0^t b(s, X(s)) \, \mathrm{d}s + \int_0^t \sigma(s, X(s)) \, \mathrm{d}W(s). \tag{3.1.7}$$

Furthermore, for all $t \in [0, \infty[$

$$E(|X(t)|^2 e^{-\alpha_t(1)}) \leqslant E(|X_0|^2) + 1. \tag{3.1.8}$$

Remark 3.1.2. *We note that by (3.1.1) the integrals on the right-hand side of (3.1.7) are well-defined.*

For the proof of the above theorem we need two lemmas.

Lemma 3.1.3. *Let $Y(t)$, $t \in [0, \infty[$, be a continuous, $\mathbb{R}_+$-valued, $(\mathcal{F}_t)$-adapted process on $(\Omega, \mathcal{F}, P)$ and γ an $(\mathcal{F}_t)$-stopping time, and let $\varepsilon \in (0, \infty)$. Set*

$$\tau_\varepsilon := \gamma \wedge \inf\{t \geqslant 0 | Y(t) \geqslant \varepsilon\}$$

(where as usual we set $\inf \emptyset = +\infty$). Then

$$P(\{\sup_{t \in [0, \gamma]} Y(t) \geqslant \varepsilon\}) \leqslant \frac{1}{\varepsilon} E(Y(\tau_\varepsilon)).$$

Proof. We have

$$\{ \sup_{t \in [0,\gamma]} Y(t) \geq \varepsilon \} = \{ Y(\tau_\varepsilon) \geq \varepsilon \}.$$

Hence the assertion follows by Chebyshev's inequality. □

The following general "localization lemma" will be crucial.

Lemma 3.1.4. *Let $n \in \mathbb{N}$ and $X^{(n)}(t)$, $t \in [0, \infty[$, be a continuous, $\mathbb{R}^d$-valued, $(\mathcal{F}_t)$-adapted process on $(\Omega, \mathcal{F}, P)$ such that $X^{(n)}(0) = X_0$ for some $\mathcal{F}_0$-measurable function $X_0 : \Omega \to \mathbb{R}^d$ and*

$$\mathrm{d}X^{(n)}(t) = b(t, X^{(n)}(t) + p^{(n)}(t)) \, \mathrm{d}t + \sigma(t, X^{(n)}(t) + p^{(n)}(t)) \, \mathrm{d}W(t), \quad t \in [0, \infty[$$

for some progressively measurable process $p^{(n)}(t)$, $t \in [0, \infty[$. For $n \in \mathbb{N}$ and $R \in [0, \infty[$ let $\tau^{(n)}(R)$ be $(\mathcal{F}_t)$-stopping times such that

(i)

$$|X^{(n)}(t)| + |p^{(n)}(t)| \leq R \quad if \quad t \in]0, \tau^{(n)}(R)] \quad P\text{-a.e.}$$

(ii)

$$\lim_{n \to \infty} E \int_0^{T \wedge \tau^{(n)}(R)} |p^{(n)}(t)| \, \mathrm{d}t = 0 \quad for \ all \ T \in [0, \infty[.$$

(iii) There exists a function $r : [0, \infty[\to [0, \infty[$ such that $\lim_{R \to \infty} r(R) = \infty$ and

$$\lim_{R \to \infty} \overline{\lim_{n \to \infty}} \, P\Big(\Big\{ \tau^{(n)}(R) \leq T, \sup_{t \in [0, \tau^{(n)}(R)]} |X^{(n)}(t)| \leq r(R) \Big\} \Big)$$

$$= 0 \ for \ all \ T \in [0, \infty[.$$

Then for every $T \in [0, \infty[$ we have

$$\sup_{t \in [0, T]} |X^{(n)}(t) - X^{(m)}(t)| \to 0 \quad in \ probability \ as \ n, m \to \infty.$$

Proof. By (3.1.1) we may assume that

$$\sup_{|x| \leq R} |b(t, x)| \leq K_t(R) \quad \text{for all } R, t \in [0, \infty[. \tag{3.1.9}$$

(Otherwise, we replace $K_t(R)$ by the maximum of $K_t(R)$ and the integrand in (3.1.1).) Fix $R \in [0, \infty[$ and define the $(\mathcal{F}_t)$-stopping times

$$\tau(R, u) := \inf\{t \geq 0 | \alpha_t(R) > u\}, \ u \in [0, \infty[.$$

Since $t \mapsto \alpha_t(R)$ is locally bounded, we have that $\tau(R, u) \uparrow \infty$ as $u \to \infty$. In particular, there exists $u(R) \in [0, \infty[$ such that

$$P(\{\tau(R, u(R)) \leq R\}) \leq \frac{1}{R}.$$

Setting $\tau(R) := \tau(R, u(R))$ we have $\tau(R) \to \infty$ in probability as $R \to \infty$ and $\alpha_{t \wedge \tau(R)}(R) \leqslant u(R)$ for all $t, R \in [0, \infty[$.

Furthermore, if we replace $\tau^{(n)}(R)$ by $\tau^{(n)}(R) \wedge \tau(R)$ for $n \in \mathbb{N}$, $R \in [0, \infty[$, then clearly assumptions (i) and (ii) above still hold. But

$$P\left(\left\{\tau^{(n)}(R) \wedge \tau(R) \leqslant T, \sup_{t \in [0, \tau^{(n)}(R) \wedge \tau(R)]} |X^{(n)}(t)| \leqslant r(R)\right\}\right)$$

$$\leqslant P\left(\left\{\tau^{(n)}(R) \leqslant T, \sup_{t \in [0, \tau^{(n)}(R)]} |X^{(n)}(t)| \leqslant r(R), \tau^{(n)}(R) \leqslant \tau(R)\right\}\right)$$

$$+ P(\{\tau(R) \leqslant T, \tau^{(n)}(R) > \tau(R)\})$$

and $\lim_{R \to \infty} P(\{\tau(R) \leqslant T\}) = 0$. So, also assumption (iii) holds when $\tau^{(n)}(R)$ is replaced by $\tau^{(n)}(R) \wedge \tau(R)$. We may thus assume that $\tau^{(n)}(R) \leqslant \tau(R)$, hence

$$\alpha_{t \wedge \tau^{(n)}(R)}(R) \leqslant u(R) \text{ for all } t, R \in [0, \infty[, n \in \mathbb{N}. \tag{3.1.10}$$

Fix $R \in [0, \infty[$ and define

$$\lambda_t^{(n)}(R) := \int_0^t |p^{(n)}(s)| K_s(R) \, ds, \quad t \in]0, \infty[, n \in \mathbb{N}. \tag{3.1.11}$$

By (3.1.10) it follows that

$$\lim_{n \to \infty} E\left(\lambda_{T \wedge \tau^{(n)}(R)}^{(n)}(R)\right) = 0 \text{ for all } R, T \in [0, \infty[. \tag{3.1.12}$$

Indeed, for all $m, n \in \mathbb{N}$

$$\int_0^{T \wedge \tau^{(n)}(R)} |p^{(n)}(t)| K_t(R) \, dt$$

$$\leqslant m \int_0^{T \wedge \tau^{(n)}(R)} |p^{(n)}(t)| \, dt + R \int_0^{T \wedge \tau(R)} 1_{]m, \infty[}(K_t(R)) K_t(R) \, dt.$$

By assumption (ii) we know that as $n \to \infty$ this converges in $L^1(\Omega, \mathcal{F}, P)$ to

$$R \int_0^{T \wedge \tau(R)} 1_{]m, \infty[}(K_t(R)) K_t(R) \, dt,$$

which in turn is dominated by $R \, \alpha_{T \wedge \tau(R)} \leqslant R \, u(R)$ and converges P-a.e. to zero as $m \to \infty$ by (3.1.5). So, (3.1.12) follows by Lebesgue's dominated convergence theorem. Let $n, m \in \mathbb{N}$ and set

$$\psi_t(R) := \exp(-2\alpha_t(R) - |X_0|), \quad t \in [0, \infty[. \tag{3.1.13}$$

Then by Itô's formula we have P-a.e. for all $t \in [0, \infty[$

$$|X^{(n)}(t) - X^{(m)}(t)|^2 \psi_t(R)$$

$$= \int_0^t \psi_s(R) \Big[2\langle X^{(n)}(s) - X^{(m)}(s), b(s, X^{(n)}(s) + p^{(n)}(s))$$

$$- b(s, X^{(m)}(s) + p^{(m)}(s))\rangle \qquad (3.1.14)$$

$$+ \|\sigma(s, X^{(n)}(s) + p^{(n)}(s)) - \sigma(s, X^{(m)}(s) + p^{(m)}(s))\|^2$$

$$- 2K_s(R)|X^{(n)}(s) - X^{(m)}(s)|^2 \Big] \, ds + M_R^{(n,m)}(t),$$

where $M_R^{(n,m)}(t)$, $t \in [0, \infty[$, is a continuous local $(\mathcal{F}_t)$-martingale with $M_R^{(n,m)}(0) = 0$. Writing

$$X^{(n)}(s) - X^{(m)}(s) = (X^{(n)}(s) + p^{(n)}(s)) - (X^{(m)}(s) + p^{(m)}(s)) - p^{(n)}(s) + p^{(m)}(s)$$

and by the weak monotonicity assumption (3.1.3), for $t \in [0, \tau^n(R) \wedge \tau^m(R)]$ the right-hand side of (3.1.14) is P-a.e. dominated by

$$\int_0^t \psi_s(R) \Big[2\langle p^{(m)}(s) - p^{(n)}(s), b(s, X^{(n)}(s) + p^{(n)}(s))$$

$$- b(s, X^{(m)}(s) + p^{(m)}(s))\rangle$$

$$+ K_s(R)|(X^{(n)}(s) - X^{(m)}(s)) + (p^{(n)}(s) - p^{(m)}(s))|^2$$

$$- 2K_s(R)|X^{(n)}(s) - X^{(m)}(s)|^2 \Big] \, ds + M_R^{(n,m)}(t)$$

$$\leqslant 2 \int_0^t \psi_s(R) \, K_s(R) \left(2|p^{(m)}(s) - p^{(n)}(s)| + |p^{(m)}(s) - p^{(n)}(s)|^2 \right) \, ds$$

$$+ M_R^{(n,m)}(t),$$

where we used (3.1.9) and assumption (i) in the last step. Since $\psi_s(R) \leqslant 1$ for all $s \in [0, \infty[$ and since for $s \in]0, \tau^{(n)}(R) \wedge \tau^{(m)}(R)]$

$$|p^{(m)}(s) - p^{(n)}(s)|^2 \leqslant 2R(|p^{(m)}(s)| + |p^{(n)}(s)|) \quad \text{P-a.e.},$$

the above implies that for $T \in [0, \infty[$ fixed and $\gamma^{(n,m)}(R) := T \wedge \tau^{(n)}(R) \wedge \tau^{(m)}(R)$ we have P-a.e. for $t \in [0, \gamma^{(n,m)}(R)]$

$$|X^{(n)}(t) - X^{(m)}(t)|^2 \psi_t(R) \leqslant 4(1 + R)(\lambda_t^{(n)}(R) + \lambda_t^{(m)}(R)) + M_R^{(n,m)}(t).$$
$$(3.1.15)$$

Hence for any $(\mathcal{F}_t)$-stopping time $\tau \leqslant \gamma^{(n,m)}(R)$ and $(\mathcal{F}_t)$-stopping times $\sigma_k \uparrow \infty$ as $k \to \infty$ so that $M_R^{(n,m)}(t \wedge \sigma_k)$, $t \in [0, \infty[$, is a martingale for all

$k \in \mathbb{N}$, we have

$$E(|X^{(n)}(\tau \wedge \sigma_k) - X^{(m)}(\tau \wedge \sigma_k)|^2 \psi_{\tau \wedge \sigma_k}(R))$$

$$\leqslant 4(1 + R)E(\lambda^{(n)}_{T \wedge \tau^{(n)}(R)}(R) + \lambda^{(m)}_{T \wedge \tau^{(m)}(R)}(R)).$$

First letting $k \to \infty$ and applying Fatou's lemma, and then using Lemma 3.1.3 we obtain that for every $\varepsilon \in]0, \infty[$

$$P(\{ \sup_{t \in [0, \gamma^{(n,m)}(R)]} (|X^{(n)}(t) - X^{(m)}(t)|^2 \psi_t(R)) > \varepsilon \})$$

$$\leqslant \frac{4(1 + R)}{\varepsilon} E(\lambda^{(n)}_{T \wedge \tau^{(n)}(R)}(R) + \lambda^{(m)}_{T \wedge \tau^{(m)}(R)}(R)).$$

Since $[0, \infty[\ni t \mapsto \psi_t(R)(\omega)$ is strictly positive, independent of $n, m \in \mathbb{N}$, and continuous, the above inequality and (3.1.12) imply that

$$\sup_{t \in [0, \gamma^{(n,m)}(R)]} |X^{(n)}(t) - X^{(m)}(t)| \to 0 \quad \text{as } n, m \to \infty$$

in P-measure. So, to prove the assertion it remains to show that given $T \in [0, \infty[$,

$$\lim_{R \to \infty} \overline{\lim_{n \to \infty}} \, P(\{\tau^{(n)}(R) \leqslant T\}) = 0. \tag{3.1.16}$$

We first observe that replacing $K_t(R)$ by $\max(K_t(R), K_t(1))$ we may assume that

$$K_t(1) \leqslant K_t(R) \text{ for all } t \in [0, \infty[, R \in [1, \infty[. \tag{3.1.17}$$

Now we proceed similarly as above, but use the assumption of weak coercivity (3.1.4) instead of the weak monotonicity (3.1.3). Let $n \in \mathbb{N}$ and $R \in [1, \infty[$. Then by Itô's formula P-a.e. for all $t \in [0, \infty[$ we have

$$|X^{(n)}(t)|^2 \psi_t(1)$$

$$= |X_0|^2 e^{-|X_0|} + \int_0^t \psi_s(1) \big[2 \langle X^{(n)}(s), b(s, X^{(n)}(s) + p^{(n)}(s)) \rangle$$

$$+ \|\sigma(s, X^{(n)}(s) + p^{(n)}(s))\|^2 - 2K_s(1)|X^{(n)}(s)|^2 \big] \, ds + M_R^{(n)}(t), \tag{3.1.18}$$

where $M_R^{(n)}(t), t \in [0, \infty[$, is a continuous local $(\mathcal{F}_t)$-martingale with $M_R^{(n)}(0) = 0$. By (3.1.4) and (3.1.9) and since $\psi_s(1) \leqslant 1$ for all $s \in [0, \infty[$ the second summand of the right-hand side of (3.1.18) is P-a.e. for all

$t \in [0, T \wedge \tau^{(n)}(R)]$ dominated by

$$\int_0^t \psi_s(1) \big[2\langle -p^{(n)}(s), b(s, X^{(n)}(s) + p^{(n)}(s)) \rangle$$
$$+ K_s(1)|X^{(n)}(s) + p^{(n)}(s)|^2 + K_s(1) - 2K_s(1)|X^{(n)}(s)|^2 \big] \, ds$$
$$\leqslant 2 \int_0^t K_s(R) |p^{(n)}(s)|(1 + |p^{(n)}(s)|) \, ds + \int_0^t e^{-2\alpha_s(1)} K_s(1) \, ds$$
$$\leqslant 2(1 + R)\lambda_t^{(n)}(R) + \int_0^{\alpha_t(1)} e^{-2s} \, ds,$$

$$(3.1.19)$$

where we used (3.1.9), (3.1.17) and assumption (i).

Again localizing $M_R^{(n)}(t)$, $t \in [0, \infty[$, from (3.1.18) and (3.1.19) we deduce that for every $(\mathcal{F}_t)$-stopping time $\tau \leqslant T \wedge \tau^{(n)}(R)$

$$E(|X^{(n)}(\tau)|^2 \psi_\tau(1)) \leqslant E(|X_0|^2 e^{-|X_0|}) + \frac{1}{2} + 2(1 + R)E(\lambda_{T \wedge \tau^{(n)}(R)}^{(n)}(R)).$$

Hence by Lemma 3.1.3 and (3.1.12) we obtain that for every $c \in]0, \infty[$

$$\lim_{c \to \infty} \sup_{R \in [0, \infty[} \varlimsup_{n \to \infty} P(\{ \sup_{t \in [0, T \wedge \tau^{(n)}(R)]} (|X^{(n)}(t)|^2 \psi_t(1)) \geqslant c \}) = 0.$$

Since $[0, \infty[\ni t \mapsto \psi_t(1)$ is strictly positive, independent of $n \in \mathbb{N}$ and continuous, and since $r(R) \to \infty$ as $R \to \infty$, we conclude that

$$\varlimsup_{R \to \infty} \varlimsup_{n \to \infty} P(\{ \sup_{t \in [0, \tau^{(n)}(R)]} |X^{(n)}(t)| \geqslant r(R), \tau^{(n)}(R) \leqslant T \})$$

$$\leqslant \lim_{R \to \infty} \sup_{\tilde{R} \in [0, \infty[} \varlimsup_{n \to \infty} P(\{ \sup_{t \in [0, T \wedge \tau^{(n)}(\tilde{R})]} |X^{(n)}(t)| \geqslant r(R) \}) = 0.$$

Hence (3.1.16) follows from assumption (iii). $\square$

Remark 3.1.5. *In our application of Lemma 3.1.4 below, assumption (iii) will be fulfilled, since the event under P will be empty for all $n \in \mathbb{N}$, $R \in [0, \infty[$. For a case where assumption (iii) is more difficult to check, we refer to [Kry99, Section 1].*

3.2. Proof of existence and uniqueness

Proof of Theorem 3.1.1. The proof is based on Euler's method. Fix $n \in \mathbb{N}$ and define the processes $X^{(n)}(t)$, $t \in [0, \infty[$, iteratively by setting

$$X^{(n)}(0) := X_0$$

and for $k \in \mathbb{N} \cup \{0\}$ and $t \in \left] \frac{k}{n}, \frac{k+1}{n} \right]$ by

$$X^{(n)}(t)$$
$$= X^{(n)}\left(\frac{k}{n}\right) + \int_{\frac{k}{n}}^{t} b\left(s, X^{(n)}\left(\frac{k}{n}\right)\right) \, ds + \int_{\frac{k}{n}}^{t} \sigma\left(s, X^{(n)}\left(\frac{k}{n}\right)\right) \, dW(s).$$

This is equivalent to

$$X^{(n)}(t) = X_0 + \int_0^t b(s, X^{(n)}(\kappa(n, s))) \, ds$$
$$+ \int_0^t \sigma(s, X^{(n)}(\kappa(n, s))) \, dW(s), \; t \in [0, \infty[, \tag{3.2.1}$$

where $\kappa(n, t) := [tn]/n$, and also to

$$X^{(n)}(t) = X_0 + \int_0^t b(s, X^{(n)}(s) + p^{(n)}(s)) \, ds$$
$$+ \int_0^t \sigma(s, X^{(n)}(s) + p^{(n)}(s)) \, dW(s), \; t \in [0, \infty[,$$

where

$$p^{(n)}(t) := X^{(n)}(\kappa(n, t)) - X^{(n)}(t)$$
$$= - \int_{\kappa(n,t)}^{t} b(s, X^{(n)}(\kappa(n, s))) \, ds$$
$$- \int_{\kappa(n,t)}^{t} \sigma(s, X^{(n)}(\kappa(n, s))) \, dW(s), \; t \in [0, \infty[.$$

Now fix $R \in [0, \infty[$ and define

$$\tau^{(n)}(R) := \inf\left\{ t \geqslant 0 \big| |X^{(n)}(t)| > \frac{R}{3} \right\}$$

and

$$r(R) := \frac{R}{4}.$$

Then clearly,

$$|p^{(n)}(t)| \leqslant \frac{2R}{3} \text{ and } |X^{(n)}(t)| \leqslant \frac{R}{3} \text{ if } t \in]0, \tau^{(n)}(R)].$$

In particular, condition (i) in Lemma 3.1.4 holds and the event in Lemma 3.1.4(iii) is empty for all $n \in \mathbb{N}$, $R \in [0, \infty[$, so this condition is satisfied. Let

$e_i, 1 \leqslant i \leqslant d$, be the canonical basis of $\mathbb{R}^d$ and $T \in [0, \infty[$. Since for $t \in [0, T]$

$$- \langle e_i, p^{(n)}(t) \rangle$$

$$= \int_{\kappa(n,t)}^t \langle e_i, b(s, X^{(n)}(\kappa(n,s))) \rangle \, ds + \int_{\kappa(n,t)}^t \langle e_i, \sigma(s, X^{(n)}(\kappa(n,s))) \, dW(s) \rangle,$$

it follows that for $\varepsilon \in]0, \infty[$ and $1 \leqslant i \leqslant d$, $t \in [0, \infty[$

$$P(\{|\langle e_i, p^{(n)}(t) \rangle| \geqslant 2\varepsilon, \, t \leqslant \tau^{(n)}(R)\})$$

$$\leqslant P\left(\left\{\int_{\kappa(n,t)}^t \sup_{|x| \leqslant R} |b(s,x)| \, ds \geqslant \varepsilon\right\}\right)$$

$$+ P\left(\left\{\sup_{\tilde{t} \in [0,t]} \left| \int_0^{\tilde{t} \wedge \tau^{(n)}(R)} 1_{[\kappa(n,t), T]}(s) \right.\right.$$

$$\left.\left. \langle e_i, \sigma(s, X^{(n)}(\kappa(n,s))) \, dW(s) \rangle \right| \geqslant \varepsilon\right\}\right)$$

and by Corollary D.0.2 the second summand is bounded by

$$\frac{3\delta}{\varepsilon} + P\left(\left\{\int_{\kappa(n,t)}^t \sup_{|x| \leqslant R} \|\sigma(t,x)\|^2 \, ds > \delta^2\right\}\right).$$

Altogether, letting first $n \to \infty$ and using (3.1.1), and then letting $\delta \to 0$ we obtain that for all $t \in [0, \infty[$

$$1_{[0, \tau_n(R)]}(t) \, p^{(n)}(t) \to 0 \text{ as } n \to \infty$$

in P-measure. Since

$$1_{[0, \tau_n(R)]}(t) \left| p^{n)}(t) \right| \leqslant \frac{2R}{3}, \, t \in [0, \infty[,$$

it follows by Lebesgue's dominated convergence theorem and Fubini's theorem that condition (ii) in Lemma 3.1.4 is also fulfilled. Now Lemma 3.1.4 and the fact that the space of continuous processes is complete with respect to locally (in $t \in [0, \infty[$) uniform convergence in probability imply that there exists a continuous, $(\mathcal{F}_t)$-adapted, $\mathbb{R}^d$-valued process $X(t)$, $t \in [0, \infty[$, such that for all $T \in [0, \infty[$

$$\sup_{t \in [0,T]} |X^{(n)}(t) - X(t)| \to 0 \text{ in } P\text{-measure as } n \to \infty. \tag{3.2.2}$$

To prove that X satisfies (3.1.6) we are going to take the limit in (3.2.1). To this end, fix $T \in [0, \infty[$ and $t \in [0, T]$. By (3.2.2) and because of the path continuity we only have to show that the right-hand side of (3.2.1) converges in P-measure to

$$X_0 + \int_0^t b(s, X(s)) \, ds + \int_0^t \sigma(s, X(s)) \, dW(s).$$

Since the convergence in (3.2.2) is uniform on $[0, T]$, by equicontinuity we have that also

$$\sup_{t \in [0,T]} |X^{(n)}(\kappa(n, t)) - X(t)| \to 0 \text{ in } P\text{-measure as } n \to \infty.$$

Hence for $Y^{(n)}(t) := X^{(n)}(\kappa(n, t))$ and a subsequence $(n_k)_{k \in \mathbb{N}}$

$$\sup_{t \in [0,T]} |Y^{(n_k)}(t) - X(t)| \to 0 \ P\text{-a.e. as } k \to \infty.$$

In particular, for $S(t) := \sup_{k \in \mathbb{N}} |Y^{(n_k)}(t)|$

$$\sup_{t \in [0,T]} S(t) < \infty \qquad P\text{-a.e.}. \tag{3.2.3}$$

For $R \in [0, \infty[$ define the $(\mathcal{F}_t)$-stopping time

$$\tau(R) := \inf\{t \in [0, T] | S(t) > R\} \wedge T.$$

By the continuity of b in $x \in \mathbb{R}^d$ and by (3.1.1)

$$\lim_{k \to \infty} \int_0^t b(s, X^{(n_k)}(\kappa(n_k, s))) \, ds = \int_0^t b(s, X(s)) \, ds \qquad P\text{-a.e. on } \{t \leqslant \tau(R)\}. \tag{3.2.4}$$

To handle the stochastic integrals we need another sequence of stopping times. For $R, N \in [0, \infty[$ define the $(\mathcal{F}_t)$-stopping time

$$\tau_N(R) := \inf\{t \in [0, T] | \int_0^t \sup_{|x| \leqslant R} \|\sigma(s, x)\|^2 \, ds > N\} \wedge \tau(R).$$

Then by the continuity of σ in $x \in \mathbb{R}^d$, (3.1.1), and Lebesgue's dominated convergence theorem

$$\lim_{k \to \infty} E\left(\int_0^{\tau_N(R)} \|\sigma(s, X^{(n_k)}(\kappa(n_k, s))) - \sigma(s, X(s))\|^2 \, ds \right) = 0,$$

hence

$$\int_0^t \sigma(s, X^{(n_k)}(\kappa(n_k, s))) \, dW(s) \to \int_0^t \sigma(s, X(s)) \, dW(s) \tag{3.2.5}$$

in P-measure on $\{t \leqslant \tau_N(R)\}$ as $k \to \infty$. By (3.1.1) for every $\omega \in \Omega$ there exists $N(\omega) \in [0, \infty[$ such that $\tau_N(R) = \tau(R)$ for all $N \geqslant N(\omega)$, so

$$\bigcup_{N \in \mathbb{N}} \{t \leqslant \tau_N(R)\} = \{t \leqslant \tau(R)\}.$$

Therefore, (3.2.5) holds on $\{t \leqslant \tau(R)\}$. But by (3.2.3) for P-a.e. $\omega \in \Omega$ there exists $R(\omega) \in [0, \infty[$ such that $\tau(R) = T$ for all $R \geqslant R(\omega)$. So, as above we conclude that (3.2.4) and (3.2.5) hold P-a.e. on Ω. This completes the proof for existence.

The uniqueness is a special case of the next proposition. So, let us prove the final statement. We have by Itô's formula for our solution X that P-a.e. for all $t \in [0, \infty[$

$$|X(t)|^2 e^{-\alpha_t(1)} = |X_0|^2 + \int_0^t e^{-\alpha_s(1)} \left[2\langle X(s), b(s, X(s)) \rangle + \|\sigma(s, X(s))\|^2 \right.$$
$$\left. - K_s(1)|X(s)|^2 \right] \, \mathrm{d}s + M(t),$$

where $M(t)$, $t \in [0, \infty[$, is a continuous local martingale with $M(0) = 0$. By the weak coercivity assumption (3.1.4) the latter is dominated by

$$|X_0|^2 + \int_0^{\alpha_t(1)} e^{-s} \, \mathrm{d}s + M(t).$$

So, again by localizing $M(t)$, $t \in [0, \infty[$, and Fatou's lemma we get

$$E(|X(t)|^2 e^{-\alpha_t(1)}) \leqslant E(|X_0|^2) + 1, \ t \in [0, \infty[.$$

$\square$

Proposition 3.2.1. *Let the assumptions of Theorem 3.1.1 apart from (3.1.4) be satisfied. Let $X_0, X_0^{(n)} : \Omega \to \mathbb{R}^d$, $n \in \mathbb{N}$, be $\mathcal{F}_0$-measurable such that*

$$P - \lim_{n \to \infty} X_0^{(n)} = X_0.$$

Let $T \in [0, \infty[$ and assume that $X(t), X^{(n)}(t)$, $t \in [0, T]$, $n \in \mathbb{N}$, be solutions of (3.1.6) (up to time T) such that $X(0) = X_0$ and $X^{(n)}(0) = X_0^{(n)}$ P-a.e. for all $n \in \mathbb{N}$. Then

$$P - \lim_{n \to \infty} \sup_{t \in [0, T]} |X^{(n)}(t) - X(t)| = 0. \tag{3.2.6}$$

Proof. By the characterization of convergence in P-measure in terms of P-a.e. convergent subsequences (cf. e.g. [Bau01]), we may assume that $X_0^{(n)} \to X_0$ as $n \to \infty$ P-a.e..

Fix $R \in [0, \infty[$ and define

$$\phi_t(R) := \exp(-\alpha_t(R) - \sup_n |X_0^{(n)}|), \ t \in [0, \infty[.$$

We note that since $|X_0| < \infty$, we have $\phi_t(R) > 0$ P-a.e. for all $t \in [0, \infty[$. Define

$$\gamma^{(n)}(R) := \inf\{t \geqslant 0 | |X^{(n)}(t)| + |X(t)| > R\} \wedge T.$$

Analogously to deriving (3.1.14) in the proof of Lemma 3.1.4 using the weak monotonicity assumption (3.1.3), we obtain that P-a.e. for all $t \in [0, T]$ and all $n \in \mathbb{N}$

$$|X^{(n)}(t \wedge \gamma^{(n)}(R)) - X(t \wedge \gamma^{(n)}(R))|^2 \phi_{t \wedge \gamma^{(n)}(R)}(R)$$

$$\leqslant |X_0^{(n)} - X_0|^2 e^{-\sup_n |X_0^{(n)}|} + m_R^{(n)}(t),$$

where $m_R^{(n)}(t)$, $t \in [0, T]$, are continuous local $(\mathcal{F}_t)$-martingales such that $m_R^{(n)}(0) = 0$. Hence localizing $m_R^{(n)}(t)$, $t \in [0, T]$, for any $(\mathcal{F}_t)$-stopping time $\tau \leqslant \gamma^{(n)}(R)$ we obtain that

$$E(|X^{(n)}(\tau) - X(\tau)|^2 \phi_\tau(R)) \leqslant E(|X_0^{(n)} - X_0|^2 e^{-\sup_n |X_0^{(n)}|}). \qquad (3.2.7)$$

Since the right-hand side of (3.2.7) converges to zero, by Lemma 3.1.3 we conclude that

$$P - \lim_{n \to \infty} \sup_{t \in [0,T]} \left(|X^{n)}(t \wedge \gamma^{(n)}(R)) - X(t \wedge \gamma^{(n)}(R))|^2 \phi_{t \wedge \gamma^{(n)}(R)}(R) \right) = 0.$$

$$(3.2.8)$$

Since P-a.e. the function $[0, \infty[\ni t \mapsto \phi_t(R)$ is continuous and strictly positive, (3.2.8) implies

$$P - \lim_{n \to \infty} \sup_{t \in [0,T]} |X^{(n)}(t \wedge \gamma^{(n)}(R)) - X(t \wedge \gamma^{(n)}(R))| = 0. \qquad (3.2.9)$$

But

$$P(\{\gamma^{(n)}(R) < T\})$$

$$\leqslant P(\{ \sup_{t \in [0,T]} |X^{(n)}(t \wedge \gamma^{(n)}(R))| + |X(t \wedge \gamma^{(n)}(R))|| \geqslant R\})$$

$$\leqslant P(\{ \sup_{t \in [0,T]} |X^{(n)}(t \wedge \gamma^{(n)}(R)) - X(t \wedge \gamma^{(n)}(R))|| \geqslant 1\})$$

$$+ P(\{2 \sup_{t \in [0,T]} |X(t)| \geqslant R - 1\}).$$

This together with (3.2.9) implies that

$$\lim_{R \to \infty} \overline{\lim_{n \to \infty}} P(\{\gamma^{(n)}(R) < T\}) = 0. \qquad (3.2.10)$$

(3.2.9) and (3.2.10) imply (3.2.6).   □

4. A Class of Stochastic Differential Equations in Banach Spaces and Applications to Stochastic Partial Differential Equations

In this chapter we will present one specific method to solve stochastic differential equations in infinite-dimensional spaces, known as the *variational approach*. The main criterion for this approach to work is that the coefficients satisfy certain monotonicity assumptions. As the main references for Subsection 4.2 we mention [RRW06] and [KR79], but also one should check the references therein.

4.1. Gelfand triples, conditions on the coefficients and examples

Let H be a separable Hilbert space with inner product $\langle\,,\,\rangle_H$ and H^* its dual. Let V be a Banach space, such that $V \subset H$ continuously and densely. Then for its dual space V^* it follows that $H^* \subset V^*$ continuously and densely. Identifying H and H^* via the Riesz isomorphism we have that

$$V \subset H \subset V^* \qquad (4.1.1)$$

continuously and densely and if $_{V^*}\langle\,,\,\rangle_V$ denotes the dualization between V^* and V (i.e. $_{V^*}\langle z, v\rangle_V := z(v)$ for $z \in V^*, v \in V$), it follows that

$$_{V^*}\langle z, v\rangle_V = \langle z, v\rangle_H \quad \text{for all } z \in H, v \in V. \qquad (4.1.2)$$

(V, H, V^*) is called a *Gelfand triple*. Note that since $H \subset V^*$ continuously and densely, also V^* is separable, hence so is V. Furthermore, $\mathcal{B}(V)$ is generated by V^* and $\mathcal{B}(H)$ by H^*. We also have by Kuratowski's theorem that $V \in \mathcal{B}(H)$, $H \in \mathcal{B}(V^*)$ and $\mathcal{B}(V) = \mathcal{B}(H) \cap V$, $\mathcal{B}(H) = \mathcal{B}(V^*) \cap H$.

Below we want to study stochastic differential equations on H of type

$$dX(t) = A(t, X(t))dt + B(t, X(t))\,dW(t) \qquad (4.1.3)$$

with $W(t)$, $t \in [0, T]$ a cylindrical Q-Wiener process with $Q = I$ on another separable Hilbert space $(U, \langle \, , \, \rangle_U)$ and with B taking values in $L_2(U, H)$ as in Chapter 2, but with A taking values in the larger space V^*.

The solution X will, however, take values in H again. In this section we give precise conditions on A and B.

Let $T \in [0, \infty[$ be fixed and let $(\Omega, \mathcal{F}, P)$ be a complete probability space with normal filtration $\mathcal{F}_t$, $t \in [0, \infty[$. Let

$$A : [0, T] \times V \times \Omega \to V^*, \ B : [0, T] \times V \times \Omega \to L_2(U, H)$$

be *progressively measurable*, i.e. for every $t \in [0, T]$, these maps restricted to $[0, t] \times V \times \Omega$ are $\mathcal{B}([0, t]) \otimes \mathcal{B}(V) \otimes \mathcal{F}_t$-measurable. As usual by writing $A(t, v)$ we mean the map $\omega \mapsto A(t, v, \omega)$. Analogously for $B(t, v)$. We impose the following conditions on A and B:

(H1) (*Hemicontinuity*) For all $u, v, x \in V$, $\omega \in \Omega$ and $t \in [0, T]$ the map

$$\mathbb{R} \ni \lambda \mapsto {}_{V^*}\langle A(t, u + \lambda v, \omega), x \rangle_V$$

is continuous.

(H2) (*Weak monotonicity*) There exists $c \in \mathbb{R}$ such that for all $u, v \in V$

$$2\,{}_{V^*}\langle A(\cdot, u) - A(\cdot, v), u - v \rangle_V + \|B(\cdot, u) - B(\cdot, v)\|^2_{L_2(U, H)}$$
$$\leqslant c\|u - v\|^2_H \text{ on } [0, T] \times \Omega.$$

(H3) (*Coercivity*) There exist $\alpha \in \,]1, \infty[$, $c_1 \in \mathbb{R}$, $c_2 \in \,]0, \infty[$ and an $(\mathcal{F}_t)$-adapted process $f \in L^1([0, T] \times \Omega, \ dt \otimes P)$ such that for all $v \in V, t \in [0, T]$

$$2\,{}_{V^*}\langle A(t, v), v \rangle_V + \|B(t, v)\|^2_{L_2(U, H)} \leqslant c_1\|v\|^2_H - c_2\|v\|^\alpha_V + f(t) \quad on \ \Omega.$$

(H4) (*Boundedness*) There exist $c_3 \in [0, \infty[$ and an $(\mathcal{F}_t)$-adapted process $g \in L^{\frac{\alpha}{\alpha - 1}}([0, T] \times \Omega, \ dt \otimes P)$ such that for all $v \in V$, $t \in [0, T]$

$$\|A(t, v)\|_{V^*} \leqslant g(t) + c_3\|v\|^{\alpha - 1}_V \quad on \ \Omega,$$

where α is as in (H3).

Remark 4.1.1.   1. *By (H3) and (H4) it follows that for all $v \in V, t \in [0, T]$*

$$\|B(t, v)\|^2_{L_2(U, H)} \leqslant c_1\|v\|^2_H + f(t) + 2\|v\|_V \, g(t) + 2c_3\|v\|^\alpha_V \quad on \ \Omega.$$

2. *Let $\omega \in \Omega$, $t \in [0, T]$. (H1) and (H2) imply that $A(t, \cdot, \omega)$ is demicontinuous, i.e.*

$$u_n \to u \text{ as } n \to \infty \text{ (strongly) in } V$$

implies

$$A(t, u_n, \omega) \to A(t, u, \omega) \text{ as } n \to \infty \text{ weakly in } V^*$$

(cf. [Zei90, Proposition 26.4])
In particular if $H = \mathbb{R}^d$, $d \in \mathbb{N}$, hence $V = V^* = \mathbb{R}^d$, then (H1) and (H2) imply that $u \mapsto A(t, u, \omega)$ is continuous from $\mathbb{R}^d$ to $\mathbb{R}^d$.

Proof. Fix $(t, \omega) \in [0, T] \times \Omega$ and set for $u \in V$

$$A(u) := A(t, u, \omega) - cu.$$

The proof will be done in four steps.
Claim 1: A is locally bounded, i.e. for all $u \in V$ there exists a neighborhood $U(u)$ such that $A(U(u))$ is a bounded subset of V^*.

Proof of Claim 1. Consider first $u := 0$. Suppose $A(U(0))$ is unbounded for all neighborhoods $U(0)$ of 0. Then there exist $u_n \in V$ such that

$$u_n \to 0 \text{ and } \|A(u_n)\|_{V^*} \to \infty \text{ as } n \to \infty.$$

Set

$$a_n := (1 + \|A(u_n)\|_{V^*} \|u_n\|_V)^{-1}.$$

Then by (H2) for all $v \in V$

$$a_n \,_{V^*}\langle A(u_n), u_n - (\pm v)\rangle_V - a_n \,_{V^*}\langle A(\pm v), u_n - (\pm v)\rangle_V \leqslant 0,$$

hence

$$\mp a_n \,_{V^*}\langle A(u_n), v\rangle_V \leqslant -a_n \,_{V^*}\langle A(u_n), u_n\rangle_V + a_n \,_{V^*}\langle A(\pm v), u_n \mp v\rangle_V$$
$$\leqslant a_n \|A(u_n)\|_{V^*} \|u_n\|_V + \|A(\pm v)\|_{V^*} \|u_n \mp v\|_V$$
$$\leqslant 1 + \|A(\pm v)\|_{V^*} \left(\sup_n \|u_n\|_V + \|v\|_V\right).$$

Consequently,

$$\sup_n |\,_{V^*}\langle a_n A(u_n), v\rangle_V| < \infty \text{ for all } v \in V.$$

Therefore, by the Banach–Steinhaus theorem

$$N := \sup_n \|a_n A(u_n)\|_{V^*} < \infty,$$

and thus for $n_0 \in \mathbb{N}$ so large that $\|u_n\| \leqslant \frac{1}{2N}$ for all $n \geqslant n_0$ we obtain

$$\|A(u_n)\|_{V^*} \leqslant a_n^{-1} N \leqslant N + \frac{1}{2}\|A(u_n)\|_{V^*},$$

i.e.

$$\|A(u_n)\|_{V^*} \leqslant 2N \text{ for all } n \geqslant n_0,$$

which is a contradiction. So, $A(U(0))$ is bounded for some neighborhood $U(0)$ of 0.

For arbitrary $u \in V$ we apply the above argument to the operator

$$A_u(v) := A(u+v), \ v \in V$$

which obviously is also hemicontinuous and weakly monotone. So, Claim 1 is proved. □

Claim 2: Let $u \in V$, $b \in V^*$ such that

$$_{V^*}\langle b - A(v), u - v \rangle_V \leqslant 0 \text{ for all } v \in V.$$

Then $A(u) = b$.

Proof of Claim 2. Let $w \in V$, $t \in]0, \infty[$ and set $v := u - tw$. Then

$$_{V^*}\langle b - A(u - tw), tw \rangle_V = {}_{V^*}\langle b - A(v), u - v \rangle_V \leqslant 0.$$

Dividing first by t and then letting $t \to 0$, by (H1) we obtain

$$_{V^*}\langle b - A(u), w \rangle_V \leqslant 0 \text{ for all } w \in V.$$

So, replacing w by $-w$, $w \in V$, we get

$$_{V^*}\langle b - A(u), w \rangle_V = 0 \text{ for all } w \in V,$$

hence $A(u) = b$. □

Claim 3: ("monotonicity trick"). Let $u_n, u \in V, n \in \mathbb{N}$, and $b \in V^*$ such that

$$u_n \to u \text{ as } n \to \infty \text{ weakly in } V,$$

$$A(u_n) \to b \text{ as } n \to \infty \text{ weakly in } V^*$$

and

$$\overline{\lim} \ _{V^*}\langle A(u_n), u_n \rangle_V \geqslant {}_{V^*}\langle b, u \rangle_V .$$

Then $A(u) = b$.

Proof of Claim 3. We have for all $v \in V$

$$_{V^*}\langle A(u_n), u_n \rangle_V - {}_{V^*}\langle A(v), u_n \rangle_V - {}_{V^*}\langle A(u_n) - A(v), v \rangle_V$$

$$= {}_{V^*}\langle A(u_n) - A(v), u_n - v \rangle_V \leqslant 0.$$

Letting $n \to \infty$ we obtain

$$_{V^*}\langle b, u \rangle_V - {}_{V^*}\langle A(v), u \rangle_V - {}_{V^*}\langle b - A(v), v \rangle_V \leqslant 0,$$

so

$$_{V^*}\langle b - A(v), u - v \rangle_V \leqslant 0 \text{ for all } v \in V.$$

Hence Claim 2 implies that $A(u) = b$. □

Claim 4: Let $u_n, u \in V$, $n \in \mathbb{N}$, such that

$$u_n \to u \text{ as } n \to \infty \text{ (strongly) in } V.$$

Then

$$A(u_n) \to A(u) \text{ as } n \to \infty \text{ weakly in } V^*.$$

Proof of Claim 4. Since $\{u_n | n \in \mathbb{N}\}$ is bounded, by Claim 1 also $\{A(u_n) | n \in \mathbb{N}\}$ is bounded in V^*. Since bounded sets in V^* are weakly compact by the Banach-Alaoglu theorem, there exists a subsequence $(n_k)_{k \in \mathbb{N}}$ and $b \in V^*$ such that $A(u_{n_k}) \to b$ as $k \to \infty$ weakly in V^*. Since $u_{n_k} \to u$ strongly in V as $k \to \infty$, we get

$$\lim_{k \to \infty} {}_{V^*}\langle A(u_{n_k}), u_{n_k}\rangle_V = {}_{V^*}\langle b, u\rangle_V .$$

Therefore, all conditions in Claim 3 are fulfilled and we can conclude that $A(u) = b$. So, for all such subsequences their weak limit is $A(u)$, hence $A(u_n) \to A(u)$ as $n \to \infty$ weakly in V^*. $\qquad \square$

Let us now discuss the above conditions. We shall solely concentrate on A and take $B \equiv 0$. The latter we do because of the following:

Exercise 4.1.2.

1. Suppose A, B satisfy (H2), (H3) above and $\tilde{A}$ is another map as A satisfying (H2), (H3). Then $A + \tilde{A}$, B satisfy (H2),(H3). Likewise, if A and $\tilde{A}$ both satisfy (H1), (H4) then so does $A + \tilde{A}$.

2. If A satisfies (H2), (H3) (with $B \equiv 0$) and for all $t \in [0, T]$, $\omega \in \Omega$, the map $u \mapsto B(t, u, \omega)$ is Lipschitz with Lipschitz constant independent of $t \in [0, T]$, $\omega \in \Omega$ then A, B satisfy (H2), (H3).

Below, we only look at A independent of $t \in [0, T]$, $\omega \in \Omega$. From here examples for A dependent on (t, ω) are then immediate.

Example 4.1.3. $V = H = V^*$ (which includes the case $H = \mathbb{R}^d$)
Clearly, since for all $v \in V$

$$2 {}_{V^*}\langle A(v), v\rangle_V \leqslant 2 {}_{V^*}\langle A(v) - A(0), v\rangle_V + \|A(0)\|_{V^*}^2 + \|v\|_V^2,$$

in the present case where $V = H = V^*$, (H2) implies (H3) with $c_1 > c_2$ and $\alpha := 2$. Furthermore, obviously, if A is Lipschitz in u then (H1)–(H4) are immediately satisfied. But for (H1)–(H3) to hold, purely local conditions (with respect to u) on A can be sufficient, as the following proposition shows.

Proposition 4.1.4. *Suppose $A : H \to H$ is Fréchet differentiable such that for some $c \in [0, \infty[$ the operator $DA(x) - cI$ $(\in L(H))$ is negative definite for all $x \in H$. Then A satisfies (H1)–(H3) (with $B \equiv 0$).*

Proof. Since A is Fréchet differentiable it is continuous, so, in particular, (H1) holds. Furthermore, for $x, y \in H$ we have

$$A(x) - A(y) = \int_0^1 \frac{d}{ds} A(y + s(x - y)) ds$$

$$= \int_0^1 DA(y + s(x - y))(x - y) ds.$$

Hence by assumption

$$\langle A(x) - A(y), x - y \rangle_H = \int_0^1 \langle DA(y + s(x - y))(x - y), x - y \rangle_H ds$$

$$\leqslant c \int_0^1 \langle x - y, x - y \rangle_H ds$$

$$= c \| x - y \|_H^2,$$

and so (H2) holds and hence (H3), as shown above. □

We again note that Proposition 4.1.4 shows that purely local conditions on A can already imply (H1)–(H3), if ($V = H = V^*$ and) $\alpha = 2$. However, the global condition (H4) then requires that A is of at most linear growth since $\alpha - 1 = 1$ if $\alpha = 2$. We also note that for $H = \mathbb{R}^1$ the conditions in Proposition 4.1.4 just mean that A is differentiable and decreasing.

If H is a space of functions, a possible and easy choice for A would be e.g. $Au = -u^3$. But then we cannot choose $H = L^2$ because A would not leave L^2 invariant. This is one motivation to look at triples $V \subset H \subset V^*$ because then we can take $V = L^p$ and $H = L^2$ and define A from V to $V^* = L^{p/(p-1)}$. Let us look at this case more precisely.

Example 4.1.5 ($L^p \subset L^2 \subset L^{p/(p-1)}$ and $A(u) := -u|u|^{p-2}$).
Let $p \in [2, \infty[$, $\Lambda \subset \mathbb{R}^d$, Λ open. Let

$$V := L^p(\Lambda) := L^p(\Lambda, \, d\xi),$$

equipped with its usual norm $\| \cdot \|_p$, and

$$H := L^2(\Lambda) := L^2(\Lambda, \, d\xi),$$

where $d\xi$ denotes Lebesgue measure on Λ. Then

$$V^* = L^{p/(p-1)}(\Lambda).$$

If $p > 2$ we assume that

$$|\Lambda| := \int_{\mathbb{R}^d} \mathbb{I}_\Lambda(\xi) \, d\xi < \infty. \tag{4.1.4}$$

Then
$$V \subset H \subset V^*,$$
or concretely
$$L^p(\Lambda) \subset L^2(\Lambda) \subset L^{p/(p-1)}(\Lambda)$$
continuously and densely. Recall that since $p > 1$, $L^p(\Lambda)$ is reflexive. Define $A : V \to V^*$ by

$$Au := -u|u|^{p-2}, \ u \in V = L^p(\Lambda).$$

Indeed, A takes values in $V^* = L^{p/(p-1)}(\Lambda)$, since

$$\int |Au(\xi)|^{p/(p-1)} \, \mathrm{d}\xi = \int |u(\xi)|^p \, \mathrm{d}\xi < \infty$$

for all $u \in L^p(\Lambda)$.

Claim: A satisfies (H1)–(H4).

Proof. Let $u, v, x \in V$. Then for $\lambda \in \mathbb{R}$

$$_{V^*}\langle A(u + \lambda v) - A(u), x \rangle_V$$
$$= \int (u(\xi)|u(\xi)|^{p-2} - (u(\xi) + \lambda v(\xi))|u(\xi) + \lambda v(\xi)|^{p-2})x(\xi) \, \mathrm{d}\xi$$
$$\leqslant \big\| u|u|^{p-2} - (u + \lambda v)|u + \lambda v|^{p-2} \big\|_{V^*} \|x\|_V$$

which converges to zero as $\lambda \to 0$ by Lebesgue's dominated convergence theorem. So, (H1) holds.
Furthermore,

$$_{V^*}\langle A(u) - A(v), u - v \rangle_V$$
$$= \int (v(\xi)|v(\xi)|^{p-2} - u(\xi)|u(\xi)|^{p-2})(u(\xi) - v(\xi)) \, \mathrm{d}\xi \leqslant 0,$$

since the map $s \mapsto s|s|^{p-2}$ is increasing on $\mathbb{R}$. Thus (H2) holds, with $c := 0$. We also have that

$$_{V^*}\langle A(v), v \rangle_V = -\int |v(\xi)|^p \, \mathrm{d}\xi = -\|v\|_V^p,$$

so (H3) holds with $\alpha := p$. In addition,

$$\|A(v)\|_{V^*} = \left(\int |v(\xi)|^p \, \mathrm{d}\xi \right)^{\frac{p-1}{p}} = \|v\|_V^{p-1}$$

so (H4) holds with $\alpha := p$ as required. $\qquad\qquad\qquad\square$

Remark 4.1.6. *In the example above we may take* $A : V := L^p(\Lambda) \to L^{\frac{p}{p-1}}(\Lambda) = V^*$ *defined by*

$$A(v) := -\Psi(v), v \in L^p(\Lambda),$$

where $\Psi : \mathbb{R} \to \mathbb{R}$ *is a fixed function satisfying properties* $(\Psi 1) - (\Psi 4)$ *specified in Example 4.1.11 below.*

Now we turn to cases where A is given by a (possibly nonlinear) partial differential operator. We shall start with the linear case; more concretely, A will be given by the classical Laplace operator

$$\Delta = \sum_{i=1}^{d} \frac{\partial^2}{\partial \xi_i^2}$$

with initial domain given by $C_0^\infty(\Lambda)$. We want to take A to be an extension of Δ to a properly chosen Banach space V so that $A : V \to V^*$ is (defined on all of V and) continuous with respect to $\|\cdot\|_V$ and $\|\cdot\|_{V^*}$. The right choice for V is the classical Sobolev space $H_0^{1,p}(\Lambda)$ for $p \in [2, \infty[$ with Dirichlet boundary conditions. So, as a preparation we need to introduce (first-order) Sobolev spaces.

Again let $\Lambda \subset \mathbb{R}^d$, Λ open, and let $C_0^\infty(\Lambda)$ denote the set of all infinitely differentiable real-valued functions on Λ with compact support. Let $p \in [1, \infty[$ and for $u \in C_0^\infty(\Lambda)$ define

$$\|u\|_{1,p} := \left(\int (|u(\xi)|^p + |\nabla u(\xi)|^p) \, d\xi \right)^{1/p}. \tag{4.1.5}$$

Then define

$$H_0^{1,p}(\Lambda) := \text{ completion of } C_0^\infty(\Lambda) \text{ with respect to } \|\cdot\|_{1,p}. \tag{4.1.6}$$

At this stage $H_0^{1,p}(\Lambda)$, called the *Sobolev space* of order 1 in $L^p(\Lambda)$ with *Dirichlet boundary conditions*, just consists of abstract objects, namly equivalence classes of $\|\cdot\|_{1,p}$-Cauchy sequences. The main point is to show that

$$H_0^{1,p}(\Lambda) \subset L^p(\Lambda), \tag{4.1.7}$$

i.e. that the unique continuous extension

$$\bar{i} : H_0^{1,p}(\Lambda) \to L^p(\Lambda)$$

of the embedding

$$i : C_0^\infty(\Lambda) \hookrightarrow L^p(\Lambda)$$

is one-to-one. To this end it suffices (in fact it is equivalent) to show that if $u_n \in C_0^\infty(\Lambda)$, $n \in \mathbb{N}$, such that

$$u_n \to 0 \quad \text{in } L^p(\Lambda)$$

and

$$\int |\nabla(u_n - u_m)(\xi)|^p \, d\xi \to 0 \text{ as } n, m \to \infty,$$

then

$$\int |\nabla(u_n(\xi))|^p \, d\xi \to 0 \text{ as } n \to \infty. \tag{4.1.8}$$

But by the completeness of $L^p(\Lambda; \mathbb{R}^d)$ there exists

$$F = (F_1, \ldots, F_d) \in L^p(\Lambda; \mathbb{R}^d)$$

such that $\nabla u_n \to F$ as $n \to \infty$ in $L^p(\Lambda; \mathbb{R}^d)$. Let $v \in C_0^\infty(\Lambda)$. Then for $1 \leqslant i \leqslant d$, integrating by parts we obtain that

$$\int v(\xi) F_i(\xi) \, d\xi = \lim_{n \to \infty} \int v(\xi) \frac{\partial}{\partial \xi_i} u_n(\xi) \, d\xi$$

$$= -\lim_{n \to \infty} \int \frac{\partial}{\partial \xi_i} v(\xi) u_n(\xi) \, d\xi$$

$$= 0.$$

Hence $F_i = 0$ $d\xi$-a.e. for all $1 \leqslant i \leqslant d$, so (4.1.8) holds.

Consider the operator

$$\nabla : C_0^\infty(\Lambda) \subset L^p(\Lambda) \to L^p(\Lambda; \mathbb{R}^d).$$

By what we have shown above, we can extend ∇ to all of $H_0^{1,p}(\Lambda)$ as follows. Let $u \in H_0^{1,p}(\Lambda)$ and let $u_n \in C_0^\infty(\Lambda)$ such that $\lim_{n \to \infty} \|u - u_n\|_{1,p} = 0$. In particular, $(\nabla u_n)_{n \in \mathbb{N}}$ is a Cauchy sequence in $L^p(\Lambda; \mathbb{R}^d)$, hence has a limit there. So, define

$$\nabla u := \lim_{n \to \infty} \nabla u_n \quad \text{in } L^p(\Lambda; \mathbb{R}^d). \tag{4.1.9}$$

By what we have shown above this limit only depends on u and not on the chosen sequence. We recall the fact that $H_0^{1,p}(\Lambda)$ is reflexive for all $p \in]1, \infty[$ (cf. [Zei90]).

Example 4.1.7 $(H_0^{1,2} \subset L^2 \subset (H_0^{1,2})^*, A = \Delta)$.
Though later we shall see that to have (H3) we have to take $p = 2$, we shall first consider for $p \in [2, \infty[$ and define

$$V := H_0^{1,p}(\Lambda), H := L^2(\Lambda),$$

so

$$V^* := H_0^{1,p}(\Lambda)^*.$$

Again we assume (4.1.4) to hold if $p > 2$. Since then $V \subset L^p(\Lambda) \subset H$, continuously and densely, identifying H with its dual we obtain the continuous and dense embeddings

$$V \subset H \subset V^*$$

or concretely

$$H_0^{1,p}(\Lambda) \subset L^2(\Lambda) \subset H_0^{1,p}(\Lambda)^*. \qquad (4.1.10)$$

Now we are going to extend Δ with initial domain $C_0^\infty(\Lambda)$ to a bounded linear operator $A : V \to V^*$. First of all we can consider Δ as an operator taking values in V^* since

$$\Delta : C_0^\infty(\Lambda) \to C_0^\infty(\Lambda) \subset L^2(\Lambda) \subset V^*.$$

Furthermore, for $u, v \in C_0^\infty(\Lambda)$ again integrating by parts we obtain

$$\big|_{V^*}\langle \Delta u, v\rangle_V\big| = |\langle \Delta u, v\rangle_H|$$

$$= \left| -\int \langle \nabla u(\xi), \nabla v(\xi)\rangle \, d\xi \right|$$

$$\leqslant \left(\int |\nabla u(\xi)|^{\frac{p}{p-1}} \, d\xi \right)^{\frac{p-1}{p}} \left(\int |\nabla v(\xi)|^p \, d\xi \right)^{\frac{1}{p}}$$

$$\leqslant \left(\int |\nabla u(\xi)|^{\frac{p}{p-1}} \, d\xi \right)^{\frac{p-1}{p}} \|v\|_{1,p}.$$

Hence for all $u \in C_0^\infty(\Lambda)$

$$\|\Delta u\|_{V^*} \leqslant \||\nabla u|\|_{\frac{p}{p-1}}. \qquad (4.1.11)$$

So, by (4.1.4) and since $\frac{p}{p-1} \leqslant 2 \leqslant p$, we get by Hölder's inequality

$$\|\Delta u\|_{V^*} \leqslant |\Lambda|^{\frac{p-2}{p}} \|u\|_{1,p} \quad \text{for all } u \in C_0^\infty(\Lambda), \qquad (4.1.12)$$

where for $p = 2$ the factor on the right is just equal to 1.
So, Δ with domain $C_0^\infty(\Lambda)$ extends (uniquely) to a bounded linear operator $A : V \to V^*$ (with domain all of V), also sometimes denoted by Δ.
Now let us check (H1)–(H4) for A.

Claim:

$$A(= \Delta) : H_0^{1,p}(\Lambda) \to \left(H_0^{1,p}(\Lambda) \right)^*$$

satisfies (H1),(H2),(H4) and provided $p = 2$, also (H3).

Proof. Since $A : V \to V^*$ is linear, (H1) is obviously satisfied. Further, if $u, v \in V$ then there exists $u_n, v_n \in C_0^\infty(\Lambda)$, $n \in \mathbb{N}$, such that $u_n \to u$, $v_n \to v$ as $n \to \infty$ in V. Hence integrating by parts we get

$$_{V^*}\langle A(u) - A(v), u - v\rangle_V = \lim_{n\to\infty} {}_{V^*}\langle \Delta u_n - \Delta v_n, u_n - v_n\rangle_V$$

$$= \lim_{n\to\infty} \langle \Delta(u_n - v_n), u_n - v_n\rangle_H$$

$$= \lim_{n\to\infty} -\int |\nabla(u_n - v_n)(\xi)|^2 \, d\xi \leqslant 0.$$

So (H2) is satisfied. Furthermore,

$$2\,_{V^*}\langle A(v), v\rangle_V = \lim_{n\to\infty} 2\langle \Delta v_n, v_n\rangle_H$$

$$= -\lim_{n\to\infty} 2 \int |\nabla v_n(\xi)|^2 \, d\xi$$

$$= -2 \int |\nabla v(\xi)|^2 \, d\xi$$

$$= 2\left(\|v\|_H^2 - \|v\|_{1,2}^2\right).$$

So (H3) is satisfied if $p = 2$ with $\alpha = 2$. Furthermore, (H4), with $\alpha = 2$ is clear by (4.1.12). □

Remark 4.1.8. *The corresponding SDE (4.1.3) then reads*

$$dX(t) = \Delta X(t) \, dt + B(t, X(t)) \, dW(t).$$

If $B \equiv 0$, this is just the classical heat equation. *If $B \not\equiv 0$, but constant, the solution is an* Ornstein–Uhlenbeck process *on H.*

Example 4.1.9 ($H_0^{1,p} \subset L^2 \subset (H_0^{1,p})^*$, $A = p$-Laplacian).
Again we take $p \in [2, \infty[$, $\Lambda \in \mathbb{R}^d$, Λ open and bounded, and $V := H_0^{1,p}(\Lambda)$, $H := L^2(\Lambda)$, so $V^* = (H_0^{1,p}(\Lambda))^*$. Define $A : H_0^{1,p}(\Lambda) \to H_0^{1,p}(\Lambda)^*$ by

$$A(u) := \operatorname{div}(|\nabla u|^{p-2}\nabla u), \quad u \in H_0^{1,p}(\Lambda);$$

more precisely, given $u \in H_0^{1,p}(\Lambda)$ for all $v \in H_0^{1,p}(\Lambda)$

$$_{V^*}\langle A(u), v\rangle_V := -\int |\nabla u(\xi)|^{p-2}\langle \nabla u(\xi), \nabla v(\xi)\rangle \, d\xi \quad \text{for all } v \in H_0^{1,p}(\Lambda).$$
$$(4.1.13)$$

A is called the *p-Laplacian*, also denoted by Δ_p. Note that $\Delta_2 = \Delta$. To show that $A : V \to V^*$ is well-defined we have to show that the right-hand side of (4.1.13) defines a linear functional in $v \in V$ which is continuous with respect to $\|\cdot\|_V = \|\cdot\|_{1,p}$. First we recall that by (4.1.9) $\nabla u \in L^p(\Lambda; \mathbb{R}^d)$ for all $u \in H_0^{1,p}(\Lambda)$. Hence by Hölder's inequality

$$\int |\nabla u(\xi)|^{p-1}|\nabla v(\xi)| \, d\xi \leqslant \left(\int |\nabla u(\xi)|^p \, d\xi\right)^{\frac{p-1}{p}} \left(\int |\nabla v(\xi)|^p \, d\xi\right)^{\frac{1}{p}}$$

$$\leqslant \|u\|_{1,p}^{p-1}\|v\|_{1,p}.$$

Since this dominates the right-hand side of (4.1.13) for all $u \in H_0^{1,p}(\Lambda)$ we have that $A(u)$ is a well-defined element of $(H_0^{1,p}(\Lambda))^*$ and that

$$\|A(u)\|_{V^*} \leqslant \|u\|_V^{p-1}.$$
$$(4.1.14)$$

Now we are going to check that A satisfies (H1)–(H4).

(H1): Let $u, v, x \in H_0^{1,p}(\Lambda)$, then by (4.1.13) we have to show for $\lambda \in \mathbb{R}$, $|\lambda| \leqslant 1$

$$\lim_{\lambda \to 0} \int \left(|\nabla(u + \lambda v)(\xi)|^{p-2} \langle \nabla(u + \lambda v)(\xi), \nabla x(\xi) \rangle \right.$$
$$\left. - |\nabla u(\xi)|^{p-2} \langle \nabla u(\xi), \nabla x(\xi) \rangle \right) d\xi = 0.$$

Since obviously the integrands converge to zero as $\lambda \to 0$ $d\xi$-a.e., we only have to find a dominating function to apply Lebesgue's dominated convergence theorem. But obviously, since $|\lambda| \leqslant 1$

$$|\nabla(u + \lambda v)(\xi)|^{p-2} |\langle \nabla(u + \lambda v)(\xi), \nabla x(\xi) \rangle|$$
$$\leqslant 2^{p-1} \left(|\nabla u(\xi)|^{p-1} + |\nabla v(\xi)|^{p-1} \right) |\nabla x(\xi)|$$

and the right-hand side is in $L^1(\Lambda)$ by Hölder's inequality as we have seen above.

(H2): Let $u, v \in H_0^{1,p}(\Lambda)$. Then by (4.1.13)

$$- {}_{V^*}\langle A(u) - A(v), u - v \rangle_V$$
$$= \int \langle |\nabla u(\xi)|^{p-2} \nabla u(\xi) - |\nabla v(\xi)|^{p-2} \nabla v(\xi), \nabla u(\xi) - \nabla v(\xi) \rangle \, d\xi$$
$$= \int \left(|\nabla u(\xi)|^p + |\nabla v(\xi)|^p - |\nabla u(\xi)|^{p-2} \langle \nabla u(\xi), \nabla v(\xi) \rangle \right.$$
$$\left. - |\nabla v(\xi)|^{p-2} \langle \nabla u(\xi), \nabla v(\xi) \rangle \right) \, d\xi$$
$$\geqslant \int \left(|\nabla u(\xi)|^p + |\nabla v(\xi)|^p - |\nabla u(\xi)|^{p-1} |\nabla v(\xi)| \right.$$
$$\left. - |\nabla v(\xi)|^{p-1} |\nabla u(\xi)| \right) \, d\xi$$
$$= \int \left(|\nabla u(\xi)|^{p-1} - |\nabla v(\xi)|^{p-1} \right) \left(|\nabla u(\xi)| - |\nabla v(\xi)| \right) \, d\xi$$
$$\geqslant 0,$$

since the map $\mathbb{R}_+ \ni s \mapsto s^{p-1}$ is increasing. Hence (H2) is shown with $c = 0$.

(H3): Because Λ is bounded by Poincaré's inequality (cf. [GT83]) there exists a constant $c = c(p, d, |\Lambda|) \in]0, \infty[$ such that

$$\int |\nabla u(\xi)|^p \, d\xi \geqslant c \int |u(\xi)|^p \, d\xi \quad \text{for all } u \in H_0^{1,p}(\Lambda). \tag{4.1.15}$$

Hence by (4.1.13) for all $u \in H_0^{1,p}(\Lambda)$

$$_{V^*}\langle A(u), u \rangle_V = - \int |\nabla u(\xi)|^p \, d\xi \leqslant - \frac{\min(1, c)}{2} \|u\|_{1,p}^p.$$

So, (H3) holds with $\alpha = p$ and $c_1 = 0$. (We note that only for (H3) have we used that Λ is bounded.)

(H4): This condition holds for A by (4.1.14) with $\alpha = p$.

Before we go on to our last example which will include the case of the porous medium equation we would like to stress the following:

Remark 4.1.10. *1. If one is given $V \subset H \subset V^*$ and $A : V \to V^*$ (e.g. as in the above examples) satisfying (H1)–(H4) (with $B \equiv 0$) one can consider a "smaller" space V_0, i.e. another reflexive separable Banach space such that*

$$V_0 \subset V$$

continuously and densely, hence (by restricting the linear functionals to V_0)

$$V^* \subset V_0^*$$

continuously and densely, so altogether

$$V_0 \subset V \subset H \subset V^* \subset V_0^*.$$

Restricting A to V_0 we see that A satisfies (H1),(H2) and (H4) with respect to the Gelfand triple

$$V_0 \subset H \subset V_0^*.$$

However, since $\|\cdot\|_{V_0}$ is up to a multiplicative constant larger than $\|\cdot\|_V$, property (H3) might no longer hold. Therefore, e.g. if one considers a map A which is given by a sum of the Laplacian (cf. Example 4.1.7) and e.g. a monomial (cf. Example 4.1.5) one cannot just take any $V_0 \subset H_0^{1,2}(\Lambda) \cap L^p(\Lambda)$, since (H3) might get lost. However, if e.g. $d \geqslant 3$ and $\frac{1}{d} + \frac{1}{p} = \frac{1}{2}$, then if Λ is bounded, by a Sobolev embedding theorem (cf. [GT83, Theorems 7.10 and 7.15]), $H_0^{1,2}(\Lambda) \subset L^p(\Lambda) \ (\subset L^{p'}(\Lambda), \ p' \in [1, p])$ continuously and densely, so one can take $V := H_0^{1,2}(\Lambda)$, $H = L^2(\Lambda)$ and can consider

$$A(u) := \Delta u - u|u|^{p'-2}, \ u \in H_0^{1,2}(\Lambda).$$

Then (H3) holds with $d := 2$. So, if $p' \in [1, 2]$, also (H4) holds with $d = 2$. The corresponding SDE (4.1.3) then reads

$$dX(t) = (\Delta X(t) - X(t)|X(t)|^{p'-2}) \, dt \quad (+B(t, X(t)) \, dW(t))$$

and is called a (stochastic) reaction diffusion equation.
In the case of the p-Laplacian, $p \in [2, \infty[$, it is even easier to take sums with monomials, since clearly $H_0^{1,p}(\Lambda) \subset L^p(\Lambda)$ continuously and densely, so

$$A(u) := div(|\nabla u|^{p-2}\nabla u) - u|u|^{p-2}, \ u \in H_0^{1,p}(\Lambda),$$

satisfies (H1)–(H4), if Λ is bounded, with respect to the Gelfand triple

$$H_0^{1,p}(\Lambda) \subset L^2(\Lambda) \subset (H_0^{1,p}(\Lambda))^*.$$

But generally, taking sums of A as above requires some care and is not always possible.

2. In all our analysis the space V^* is only used as a tool. Eventually, since the solutions to our SDE (4.1.3) will take values in H, V^* will be of no relevance. Therefore, no further information about V^* such as its explicit representation (e.g. as a space of Schwartz distributions) is necessary.

Example 4.1.11. $[L^p \subset (H_0^{1,2})^* \subset (L^p)^*$, $A = $ porous medium operator]
As references for this example we refer e.g. to [Aro86], [DPRLRW06], [RRW06].
Let $\Lambda \subset \mathbb{R}^d$, Λ open and bounded, $p \in [2, \infty[$ and

$$V := L^p(\Lambda), H := (H_0^{1,2}(\Lambda))^*.$$

Since Λ is bounded we have by Poincaré's inequality (4.1.15) that for some constant $c = c(2, d, |\Lambda|) > 0$

$$\|u\|_{1,2} \geqslant \|u\|_{H_0^{1,2}} := \left(\int |\nabla u(\xi)|^2 \, d\xi \right)^{\frac{1}{2}}$$
$$\geqslant \left(\frac{\min(1,c)}{2} \right)^{\frac{1}{2}} \|u\|_{1,2} \quad \text{for all } u \in H_0^{1,2}(\Lambda). \tag{4.1.16}$$

So, we can (and will do so below) consider $H_0^{1,2}(\Lambda)$ with norm $\|\cdot\|_{H_0^{1,2}}$ and corresponding scalar product

$$\langle u, v \rangle_{H_0^{1,2}} := \int \langle \nabla u(\xi), \nabla v(\xi) \rangle \, d\xi, \ u, v \in H_0^{1,2}(\Lambda).$$

Since $H_0^{1,2}(\Lambda) \subset L^2(\Lambda)$ continuously and densely, so is

$$H_0^{1,2}(\Lambda) \subset L^{\frac{p}{p-1}}(\Lambda).$$

Hence

$$L^p(\Lambda) \equiv \left(L^{\frac{p}{p-1}}(\Lambda) \right)^* \subset (H_0^{1,2}(\Lambda))^* = H,$$

continuously and densely. Now we would like to identify H with its dual $H^* = H_0^{1,2}(\Lambda)$ via the corresponding Riesz isomorphism $R : H \to H^*$ defined by $Rx := \langle x, \cdot \rangle_H$, $x \in H$. Let us calculate the latter.

Lemma 4.1.12. The map $\Delta : H_0^{1,2}(\Lambda) \to (H_0^{1,2}(\Lambda))^* = H$ (defined by (4.1.13) for $p = 2$) is an isometric isomorphism. In particular,

$$\langle \Delta u, \Delta v \rangle_H = \langle u, v \rangle_{H_0^{1,2}} \quad \text{for all } u, v \in H_0^{1,2}(\Lambda). \tag{4.1.17}$$

Furthermore, $(-\Delta)^{-1} : H \to H^* = H_0^{1,2}(\Lambda)$ *is the Riesz isomorphism for* H, *i.e. for every* $x \in H$

$$\langle x, \cdot \rangle_H = {}_{H_0^{1,2}}\langle (-\Delta)^{-1}x, \cdot \rangle_H. \tag{4.1.18}$$

Proof. Let $u \in H_0^{1,2}(\Lambda)$. Since by (4.1.13) for all $v \in H_0^{1,2}(\Lambda)$

$$_H\langle -\Delta u, v \rangle_{H_0^{1,2}} = \int \langle \nabla u(\xi), \nabla v(\xi) \rangle \, d\xi = \langle u, v \rangle_{H_0^{1,2}}, \tag{4.1.19}$$

it follows that $-\Delta : H_0^{1,2}(\Lambda) \to H$ is just the Riesz isomorphism for $H_0^{1,2}(\Lambda)$ and the first part of the assertion including (4.1.17) follows. To prove the last part, fix $x \in H$. Then by (4.1.17) and (4.1.19) for all $y \in H$

$$\langle x, y \rangle_H = \langle (-\Delta)^{-1}x, (-\Delta)^{-1}y \rangle_{H_0^{1,2}} = {}_H\langle x, (-\Delta)^{-1}y \rangle_{H_0^{1,2}}.$$

$\square$

Now we identify H with its dual H^* by the Riesz map $(-\Delta)^{-1} : H \to H^*$, so $H \equiv H^*$ in this sense, hence

$$V = L^p(\Lambda) \subset H \subset (L^p(\Lambda))^* = V^* \tag{4.1.20}$$

continuously and densely.

Lemma 4.1.13. *The map*

$$\Delta : H_0^{1,2}(\Lambda) \to (L^p(\Lambda))^*$$

extends to a linear isometry

$$\Delta : L^{\frac{p}{p-1}}(\Lambda) \to (L^p(\Lambda))^* = V^*$$

and for all $u \in L^{\frac{p}{p-1}}(\Lambda)$, $v \in L^p(\Lambda)$

$$_{V^*}\langle -\Delta u, v \rangle_V = {}_{L^{\frac{p}{p-1}}}\langle u, v \rangle_{L^p} = \int u(\xi)v(\xi) \, d\xi. \tag{4.1.21}$$

Remark 4.1.14. *One can prove that this isometry is in fact surjective, hence*

$$(L^p(\Lambda))^* = \Delta(L^{\frac{p}{p-1}}) \neq L^{\frac{p}{p-1}}.$$

We shall not use this below, but it shows that the embedding (4.1.20) has to be handled with care taking always into account that H *is identified with* H^* *by* $(-\Delta)^{-1} : H \to H^*$ *giving rise to a different dualization between* $L^p(\Lambda)$ *and* $(L^p(\Lambda))^*$. *In particular, for all* $x \in H$, $v \in L^p(\Lambda)$

$$_{(L^p)^*}\langle x, v \rangle_{L^p} = \langle x, v \rangle_H$$

$$\left(\neq {}_{L^{\frac{p}{p-1}}}\langle x, v \rangle_{L^p} = \int x(\xi)v(\xi) \, d\xi \quad \text{provided } x \in L^{\frac{p}{p-1}} \right).$$

Proof of Lemma 4.1.13. Let $u \in H_0^{1,2}(\Lambda)$. Then since $\Delta u \in H$, by (4.1.2) and (4.1.18) we obtain that for all $v \in V$

$$_{V^*}\langle \Delta u, v \rangle_V = \langle \Delta u, v \rangle_H = -_{H_0^{1,2}}\langle u, v \rangle_H = -\langle u, v \rangle_{L^2} \qquad (4.1.22)$$

since $v \in V \subset L^2(\Lambda)$. Therefore,

$$\|\Delta u\|_{V^*} \leqslant \|u\|_{\frac{p}{p-1}}.$$

So, Δ extends to a continuous linear map

$$\Delta : L^{\frac{p}{p-1}}(\Lambda) \to V^*$$

such that (4.1.22) holds for all $u \in L^{\frac{p}{p-1}}(\Lambda)$, i.e. (4.1.21) is proved. So, applying it to $u \in L^{\frac{p}{p-1}}(\Lambda)$ and

$$v := -\|u\|_q^{-\frac{q}{p}} u|u|^{q-2} \in L^p(\Lambda),$$

where $q := \frac{p}{p-1}$, by (4.1.21) we obtain that

$$_{V^*}\langle \Delta u, v \rangle_V = \|u\|_{\frac{p}{p-1}}$$

and $\|v\|_p = 1$, so $\|\Delta u\|_{V^*} = \|u\|_{\frac{p}{p-1}}$ and the assertion is completely proved. $\square$

Now we want to define the "porous medium operator A". So, let $\Psi : \mathbb{R} \to \mathbb{R}$ be a function having the following properties:

(Ψ1) Ψ is continuous.

(Ψ2) For all $s, t \in \mathbb{R}$
$$(t - s)(\Psi(t) - \Psi(s)) \geqslant 0.$$

(Ψ3) There exist $p \in [2, \infty[$, $a \in {]}0, \infty[$, $c \in [0, \infty[$ such that for all $s \in \mathbb{R}$
$$s\Psi(s) \geqslant a|s|^p - c.$$

(Ψ4) There exist $c_3, c_4 \in {]}0, \infty[$ such that for all $s \in \mathbb{R}$
$$|\Psi(s)| \leqslant c_4 + c_3|s|^{p-1},$$

where p is as in (Ψ3).

We note that (Ψ4) implies that

$$\Psi(v) \in L^{\frac{p}{p-1}}(\Lambda) \quad \text{for all } v \in L^p(\Lambda). \qquad (4.1.23)$$

Now we can define the *porous medium operator* $A : L^p(\Lambda) = V \to V^* = (L^p(\Lambda))^*$ by

$$A(u) := \Delta\Psi(u), \quad u \in L^p(\Lambda). \qquad (4.1.24)$$

Note that by (4.1.21) and Lemma 4.1.13 the operator A is well-defined. Now let us check (H1)–(H4).

(H1): Let $u, v, x \in V = L^p(\Lambda)$ and $\lambda \in \mathbb{R}$. Then by (4.1.21)

$$
\begin{aligned}
{}_{V^*}\langle A(u + \lambda v), x \rangle_V &= {}_{V^*}\langle \Delta \Psi(u + \lambda v), x \rangle_V \\
&= -\int \Psi(u(\xi) + \lambda v(\xi)) x(\xi) \, d\xi.
\end{aligned}
\tag{4.1.25}
$$

By (Ψ4) for $|\lambda| \leqslant 1$ the integrand in the right-hand side of (4.1.25) is bounded by

$$
[c_4 + c_3 2^{p-1}(|u|^{p-1} + |v|^{p-1})]|x|
$$

which by Hölder's inequality is in $L^1(\Lambda)$. So, (H1) follows by (Ψ1) and Lebesgue's dominated convergence theorem.

(H2): Let $u, v \in V = L^p(\Lambda)$. Then by (4.1.21)

$$
\begin{aligned}
{}_{V^*}\langle A(u) - A(v), u - v \rangle_V &= {}_{V^*}\langle \Delta(\Psi(u) - \Psi(v)), u - v \rangle_V \\
&= -\int [\Psi(u(\xi)) - \Psi(v(\xi))](u(\xi) - v(\xi)) \, d\xi \\
&\leqslant 0,
\end{aligned}
$$

where we used (Ψ2) in the last step.

(H3): Let $v \in L^p(\Lambda) = V$. Then by (4.1.21) and (Ψ3)

$$
\begin{aligned}
{}_{V^*}\langle A(v), v \rangle_V &= -\int \Psi(v(\xi)) v(\xi) \, d\xi \\
&\leqslant \int (-a|v(\xi)|^p + c) \, d\xi.
\end{aligned}
$$

Hence (H3) is satisfied with $c_1 := 0$, $c_2 := 2a$, $\alpha = p$ and $f(t) = 2c|\Lambda|$.

(H4): Let $v \in L^p(\Lambda) = V$. Then by Lemma 4.1.13 and (Ψ4)

$$
\begin{aligned}
\|A(v)\|_{V^*} &= \|\Delta \Psi(v)\|_{V^*} \\
&= \|\Psi(v)\|_{L^{\frac{p}{p-1}}} \\
&\leqslant c_4 |\Lambda|^{\frac{p-1}{p}} + c_3 \left(\int |v(\xi)|^p \, d\xi \right)^{\frac{p-1}{p}} \\
&= c_4 |\Lambda|^{\frac{p-1}{p}} + c_3 \|v\|_V^{p-1},
\end{aligned}
$$

so (H4) holds with $\alpha = p$.

Remark 4.1.15. 1. *For $p \in [2, \infty[$ and $\Psi(s) := s|s|^{p-2}$ we have*

$$
A(v) = \Delta(v|v|^{p-2}), \quad v \in L^p(\Lambda),
$$

which is the non-linear operator appearing in the classical porous medium equation, i.e.

$$\frac{\partial X(t)}{\partial t} = \Delta(X(t)|X(t)|^{p-2}), \quad X(0,\cdot) = X_0,$$

whose solution describes the time evolution of the density $X(t)$ of a substance in a porous medium (cf. e.g. [Aro86]).

2. *Let $\Psi : \mathbb{R} \to \mathbb{R}$ be given such that $(\Psi 1)$–$(\Psi 4)$ are satisfied with some $p \in]1,\infty[$ (in $(\Psi 3)$, $(\Psi 4)$). One can see that the above assumptions that Λ is bounded and $p \geqslant 2$, can be avoided. But p then depends on the dimension of the underlying space $\mathbb{R}^d$. Let us assume first that $d \geqslant 3$. We distinguish two cases:*

Case 1. $|\Lambda| = \infty$ *and* $p := \frac{2d}{d+2}$, $c = c_4 = 0$, *where c, c_4 are the constants in $(\Psi 3)$ and in $(\Psi 4)$ respectively.*

Case 2. $|\Lambda| < \infty$ *and* $p \in \left[\frac{2d}{d+2}, \infty\right[$.

By the Sobolev embedding theorem (cf. [GT83, Theorem 7.10]) we have

$$H_0^{1,2}(\Lambda) \subset L^{\frac{2d}{d-2}}(\Lambda)$$

continuously and densely, and

$$\|u\|_{\frac{2d}{d-2}} \leqslant \frac{2(d-1)}{\sqrt{d(d-2)}}\|u\|_{H_0^{1,2}} \text{ for all } u \in H_0^{1,2}(\Lambda).$$

In Case 1 we have $\frac{2d}{d-2} = \frac{p}{p-1}$ and in Case 2 (hence in both cases)

$$\frac{2d}{d-2} \geqslant \frac{p}{p-1}$$

and thus

$$H_0^{1,2}(\Lambda) \subset L^{\frac{p}{p-1}}(\Lambda)$$

densely and for some $c_0 \in]0,\infty[$

$$\|u\|_{\frac{p}{p-1}} \leqslant c_0\|u\|_{H_0^{1,2}} \text{ for all } u \in H_0^{1,2}(\Lambda).$$

Now the above arguments generalize to both Cases 1 and 2, i.e. for the Gelfand triple

$$V := L^p(\Lambda) \subset H := (H_0^{1,2}(\Lambda))^* \subset (L^p(\Lambda))^*$$

the operator

$$A : L^p(\Lambda) =: V \to V^* = (L^p(\Lambda))^*$$

defined in (4.1.24), satisfies (H1)–(H4).

We note that in Case 1 the norm $\|\cdot\|_{H_0^{1,2}}$ defined in (4.1.16) is in general not equivalent to $\|\cdot\|_{1,2}$, because the Poincare inequality does not hold. So, $H_0^{1,2}(\Lambda))^*$ as a dual to the normed vector spaces $(H_0^1(\Lambda), \|\cdot\|_{H_0^{1,2}})$ is complete. In particular, if $(3 \leqslant d) \leqslant 6$, we may take $p = \frac{3}{2}$ and

$$\Psi(s) := \text{sign}(s)\sqrt{|s|}, s \in \mathbb{R}.$$

For Λ bounded the above extends, of course, also to the case $d = 1, 2$ where even stronger Sobolev embeddings hold (cf. [GT83, Theorems 7.10 and 7.15]).

4.2. The main result and an Itô formula

Consider the general situation described at the beginning of the previous section. So, we have a Gelfand triple

$$V \subset H \subset V^*$$

and maps

$$A : [0, T] \times V \times \Omega \to V^*, \quad B : [0, T] \times V \times \Omega \to L_2(U, H)$$

as specified there, satisfying (H1)–(H4), and consider the stochastic differential equation

$$dX(t) = A(t, X(t))\, dt + B(t, X(t))\, dW(t) \tag{4.2.1}$$

on H with $W(t)$, $t \in [0, T]$, a cylindrical Q-Wiener process with $Q := I$ taking values in another separable Hilbert space $(U, \langle\ ,\ \rangle_U)$ and being defined on a complete probability space $(\Omega, \mathcal{F}, P)$ with normal filtration $\mathcal{F}_t$, $t \in [0, T]$.

Before we formulate our main existence and uniqueness result for solutions of (4.2.1) we have to define what we mean by "solution".

Definition 4.2.1. A continuous H-valued $(\mathcal{F}_t)$-adapted process $(X(t))_{t \in [0,T]}$ is called a *solution of* (4.2.1), if for its $dt \otimes P$-equivalence class $\hat{X}$ we have $\hat{X} \in L^\alpha([0, T] \times \Omega,\ dt \otimes P; V) \cap L^2([0, T] \times \Omega,\ dt \otimes P; H)$ with α as in (H3) and P-a.s.

$$X(t) = X(0) + \int_0^t A(s, \bar{X}(s))ds + \int_0^t B(s, \bar{X}(s))\, dW(s), \quad t \in [0, T], \tag{4.2.2}$$

where $\bar{X}$ is any V-valued progressively measurable $dt \otimes P$-version of $\hat{X}$.

Remark 4.2.2.

1. *The existence of the special version $\bar{X}$ above follows from Exercise 4.2.3 below. Furthermore, for technical reasons in Definition 4.2.1 and below we consider all processes initially as V^*-valued, hence by $dt \otimes P$-equivalence classes we always mean classes of V^*-valued processes.*

2. *The integral with respect to ds in (4.2.2) is initially a V^*-valued Bochner integral which turns out to be in fact H-valued.*

3. *Solutions in the sense of Definition 4.2.1 are often called* variational solutions *in the literature. There are various other notions of solutions for stochastic (partial) differential equations. We recall the definition of (probabilistically) weak and strong solutions in Appendix E below. The notions of analytically weak and strong solutions as well as the notion of mild solutions and their relations are recalled in Appendix F below.*

Exercise 4.2.3.

1. *Let $B_1^{V^*}$ denote the closed unit ball in V^*. Since $B_1^{V^*} \cap H \neq \emptyset$, it has a countable subset $\{l_i | i \in \mathbb{N}\}$, which is dense in $B_1^{V^*} \cap H$ with respect to H-norm.*
   Define $\Theta : H \to [0, \infty]$ by

$$\Theta(h) := \sup_{i \in \mathbb{N}} |\langle l_i, h \rangle_H|, \quad h \in H.$$

   Then Θ is lower semicontinuous on H, hence $\mathcal{B}(H)$-measurable. Since $_{V^*}\langle l_i, v \rangle_V = \langle l_i, v \rangle_H, i \in \mathbb{N}, v \in V$, we have

$$\Theta(v) = \|v\|_V \quad \text{for all } v \in V,$$

   and furthermore (by the reflexivity of V)

$$\{\Theta < \infty\} = V.$$

2. *Let $X : [0, T] \times \Omega \to H$ be any progressively measurable (i.e. $\mathcal{B}([0,t]) \otimes \mathcal{F}_t/\mathcal{B}(H)$-measurable for all $t \in [0, T]$) $dt \otimes P$-version of $\hat{X} \in L^\alpha([0, T] \times \Omega, dt \otimes P; V), \alpha \in (0, \infty)$. Then*

$$\bar{X} := \mathbb{I}_{\{\Theta \circ X < \infty\}} X$$

   is a V-valued progressively measurable (i.e. $\mathcal{B}([0,t]) \otimes \mathcal{F}_t/\mathcal{B}(V)$-measurable) $dt \otimes P$-version of $\hat{X}$.

3. *Both $A(\cdot, \bar{X})$ and $B(\cdot, \bar{X})$ are V-valued respectively $L_2(U, H)$-valued progressively measurable processes.*

Now the main result (cf. [KR79]):

Theorem 4.2.4. *Let A, B above satisfy (H1)–(H4) and let $X_0 \in L^2(\Omega, \mathcal{F}_0, P; H)$. Then there exists a unique solution X to (4.2.1) in the sense of Definition 4.2.1. Moreover,*

$$E(\sup_{t \in [0,T]} \|X(t)\|_H^2) < \infty. \tag{4.2.3}$$

The proof of Theorem 4.2.4 strongly depends on the following Itô formula, from [KR79, Theorem I.3.1], which we shall prove here first. The presentation of its proof and that of Theorem 4.2.4 is an extended adaptation of those in [RRW06].

Theorem 4.2.5. *Let $X_0 \in L^2(\Omega, \mathcal{F}_0, P; H)$ and $Y \in L^{\frac{\alpha}{\alpha-1}}([0,T] \times \Omega, \, dt \otimes P; V^*)$, $Z \in L^2([0,T] \times \Omega, \, dt \otimes P; L_2(U,H))$, both progressively measurable. Define the continuous V^*-valued process*

$$X(t) := X_0 + \int_0^t Y(s)ds + \int_0^t Z(s) \, dW(s), \, t \in [0,T].$$

If for its $dt \otimes P$-equivalence class $\hat{X}$ we have $\hat{X} \in L^\alpha([0,T] \times \Omega, \, dt \otimes P, V)$ with α as in (H3), then X is an H-valued continuous $(\mathcal{F}_t)$-adapted process,

$$E\left(\sup_{t \in [0,T]} \|X(t)\|_H^2\right) < \infty$$

and the following Itô-formula holds for the square of its H-norm P-a.s.

$$\|X(t)\|_H^2 = \|X_0\|_H^2 + \int_0^t \left(2 \, _{V^*}\langle Y(s), \bar{X}(s)\rangle_V + \|Z(s)\|_{L_2(U,H)}^2\right) ds$$

$$+ 2\int_0^t \langle X(s), Z(s) \, dW(s)\rangle_H \quad \text{for all } t \in [0,T] \tag{4.2.4}$$

for any V-valued progressively measurable $dt \otimes P$-version $\bar{X}$ of $\hat{X}$.

As in [KR79] for the proof of Theorem 4.2.5 we need the following lemma about piecewise constant approximations based on an argument due to [Doo53]. For abbreviation below we set

$$K := L^\alpha([0,T] \times \Omega, dt \otimes P; V). \tag{4.2.5}$$

Lemma 4.2.6. *Let $X : [0,T] \times \Omega \to V^*$ be $\mathcal{B}([0,T]) \otimes \mathcal{F}/\mathcal{B}(V^*)$-measurable such that for its $dt \otimes P$-equivalence class $\hat{X}$ we have $\hat{X} \in K$. Then there exists a sequence of partitions $I_l := \{0 = t_0^l < t_1^l < \cdots < t_{k_l}^l = T\}$ such that $I_l \subset I_{l+1}$ and $\delta(I_l) := \max_i(t_i^l - t_{i-1}^l) \to 0$ as $l \to \infty$, $X(t_i^l) \in V$ P-a.e. for all $l \in \mathbb{N}, 1 \leq i \leq k_l - 1$, and for*

$$\bar{X}^l := \sum_{i=2}^{k_l} 1_{[t^l_{i-1}, t^l_i[} X(t^l_{i-1}), \quad \tilde{X}^l := \sum_{i=1}^{k_l-1} 1_{[t^l_{i-1}, t^l_i[} X(t^l_i), \quad l \in \mathbb{N},$$

we have $\bar{X}^l$, $\tilde{X}^l$ are ($dt \otimes P$-versions of elements) in K such that

$$\lim_{l \to \infty} \{\|\hat{X} - \bar{X}^l\|_K + \|\hat{X} - \tilde{X}^l\|_K\} = 0.$$

Proof. For simplicity we assume that $T = 1$ and let $\bar{X} : [0,1] \times \Omega \to V$ be a $dt \otimes P$-version of $\hat{X}$. We extend $\bar{X}$ to $\mathbb{R} \times \Omega$ by setting $\bar{X} = 0$ on $[0,1]^c \times \Omega$. There exists $\Omega' \in \mathcal{F}$ with full probability such that for every $\omega \in \Omega'$ there exists a sequence $(f_n)_{n \in \mathbb{N}} \subset C(\mathbb{R}; V)$ with compact support such that

$$\int_{\mathbb{R}} \|f_n(s) - \bar{X}(s,\omega)\|_V^\alpha \, ds \leq \frac{1}{2n}, \quad n \in \mathbb{N}.$$

Thus, for every $n \in \mathbb{N}$,

$$\limsup_{\delta \to 0} \int_{\mathbb{R}} \|\bar{X}(\delta + s, \omega) - \bar{X}(s,\omega)\|_V^\alpha \, ds$$

$$\leq 3^{\alpha-1} \limsup_{\delta \to 0} \int_{\mathbb{R}} \left[\|\bar{X}(\delta + s, \omega) - f_n(\delta + s)\|_V^\alpha + \|\bar{X}(s,\omega) - f_n(s)\|_V^\alpha\right] ds$$

$$\leq \frac{3^{\alpha-1}}{n}, \quad n \in \mathbb{N}.$$

Here we used that since each f_n is uniformly continuous, by Lebesgue's dominated convergence theorem we have that for all $n \in \mathbb{N}$

$$\lim_{\delta \to 0} \int_{\mathbb{R}} \|f_n(\delta + s) - f_n(s)\|_V^\alpha \, ds = 0.$$

Letting $n \to \infty$ we obtain

$$\lim_{\delta \to 0} \int_{\mathbb{R}} \|\bar{X}(\delta + s, \omega) - \bar{X}(s,\omega)\|_V^\alpha \, ds = 0, \quad \omega \in \Omega'. \tag{4.2.6}$$

Now, given $t \in \mathbb{R}$, let $[t]$ denote the largest integer $\leq t$. Let $\gamma_n(t) := 2^{-n}[2^n t]$, $n \in \mathbb{N}$, that is, $\gamma_n(t)$ is the largest number of the form $\frac{k}{2^n}$, $k \in \mathbb{Z}$, below t. Shifting the integral in (4.2.6) by t and taking $\delta = \gamma_n(t) - t$ we obtain

$$\lim_{n \to \infty} \int_{\mathbb{R}} \|\bar{X}(\gamma_n(t) + s) - \bar{X}(t + s)\|_V^\alpha \, ds = 0 \quad \text{on } \Omega'.$$

Moreover,

$$\int_0^1 \|\bar{X}(\gamma_n(t) + s) - \bar{X}(t + s)\|_V^\alpha \, ds$$

$$\leq 1_{[-2,2]}(t) 2^{\alpha-1} \int_{\mathbb{R}} \left[\|\bar{X}(\gamma_n(t) + s)\|_V^\alpha + \|\bar{X}(t + s)\|_V^\alpha \right] ds$$

$$= 2^\alpha 1_{[-2,2]}(t) \int_0^1 \|\bar{X}(s)\|_V^\alpha \, ds \text{ on } \Omega'.$$

So, by Lebesgue's dominated convergence theorem, we obtain that

$$0 = \lim_{n \to \infty} E \int_{\mathbb{R}} dt \int_0^1 \|\bar{X}(\gamma_n(t) + s) - \bar{X}(t + s)\|_V^\alpha \, ds$$

$$\geq \lim_{n \to \infty} E \int_0^1 ds \int_0^1 \|\bar{X}(\gamma_n(t - s) + s) - \bar{X}(t)\|_V^\alpha \, dt. \tag{4.2.7}$$

Given $s \in [0, 1)$ and $n \in \mathbb{N}$, let the partition $I_n(s)$ be defined by

$$t_0^n(s) := 0, \ t_i^n(s) := \left(s - \frac{[2^n s]}{2^n} \right) + \frac{i-1}{2^n}, \ 1 \leq i \leq 2^n, \ t_{2^n+1}^n(s) := 1.$$

Then, for $t \in [t_{i-1}^n(s), t_i^n(s)[$, $1 \leqslant i \leqslant 2^n + 1$, one has $t - s \in [2^{-n}(i - [2^n s] - 2), 2^{-n}(i - [2^n s] - 1)[$ and hence,

$$\gamma_n(t - s) + s = \left[2^{-n}(i - [2^n s] - 2) + s \right]^+ = t_{i-1}^n(s), \ 1 \leq i \leq 2^n + 1.$$

Therefore, (4.2.7) implies

$$\lim_{n \to \infty} E \int_0^1 ds \int_0^1 \|\bar{X}(t) - \bar{X}^{n,s}(t)\|_V^\alpha \, dt = 0,$$

where $\bar{X}^{n,s}$ is the process defined as $\bar{X}^l$ for the partition $I_n(s)$ but with $X(t_{i-1}^l(s))$ replaced by $\bar{X}(t_{i-1}^l(s))$. Similarly, the same holds for $\tilde{X}^{n,s}$ in place of $\bar{X}^{n,s}$ by using $\tilde{\gamma}_n := \gamma_n + 2^{-n}$ instead of γ_n, where $\tilde{X}^{n,s}$ is defined as $\tilde{X}^l$ for the partition $I_n(s)$ but with $X(t_i^l(s))$ replaced by $\bar{X}(t_i^l(s))$. Hence, there exist a subsequence $n_k \to \infty$ and a ds-zero set $N_1 \in \mathcal{B}([0, 1])$ such that

$$\lim_{k \to \infty} E \int_0^1 \left\{ \|\bar{X}(t) - \bar{X}^{n_k,s}(t)\|_V^\alpha + \|\bar{X}(t) - \tilde{X}^{n_k,s}(t)\|_V^\alpha \right\} dt = 0, \quad s \in [0, 1] \setminus N_1.$$

Since for $1 \leq i \leq 2^n$ the maps $s \mapsto t_i^n(s)$ are piecewise C^1-diffeomorphisms, the image measures of ds under these maps are absolutely continuous with respect to ds. Therefore, since $\bar{X} = X \ ds \otimes P$-a.e., there exists a ds-zero set $N_2 \in \mathcal{B}([0, 1])$ such that

$$\bar{X}(t_i^n(s)) = X(t_i^n(s)) \quad P\text{-a.e. for all } s \in [0,1] \setminus N_2, 1 \le i \le 2^n.$$

Since for any $s \in [0,1] \setminus (N_1 \cup N_2)$ one has $E\big(\|\bar{X}(t_i^n(s))\|_V^\alpha\big) < \infty$, the map

$$[0,1] \times \Omega \ni (s,\omega) \mapsto X(t_i^n(s),\omega) \in V$$

is once again (a $dt \otimes P$-version of an element) in K. Therefore, fixing $s \in [0,1] \setminus (N_1 \cup N_2)$, the sequence of the corresponding partitions $I_{n_l}(s), l \ge 1$, has all properties of the assertion. $\qquad\square$

Remark 4.2.7. *As follows from the above proof all the partition points $t_i^l, l \ge 1, 1 \le i \le k_l - 1$, in the assertion of Lemma 4.2.6 can be chosen outside an a priori given Lebesgue zero set in $[0,T]$ instead of N_2 above.*

Proof of Theorem 4.2.5. Since $M(t) := \int_0^t Z(s) \, dW(s)$, $t \in [0,T]$, is already a continuous martingale on H and since $Y \in K^* = L^{\alpha/(\alpha-1)}([0,T] \times \Omega \to V^*; dt \otimes P)$ is progressively measurable, $\int_0^t Y(s) \, ds$ is a continuous adapted process on V^*. Thus, X is a continuous adapted process on V^*, hence $\mathcal{B}([0,T]) \otimes \mathcal{F}/\mathcal{B}(V^*)$-measurable.

   Claim (a):

$$\|X(t)\|_H^2 = \|X(s)\|_H^2 + 2 \int_s^t {}_{V^*}\langle Y(r), X(t)\rangle_V \, dr + 2\langle X(s), M(t) - M(s)\rangle_H$$
$$+ \|M(t) - M(s)\|_H^2 - \|X(t) - X(s) - M(t) + M(s)\|_H^2$$
$$\tag{4.2.8}$$

holds for all $t > s$ such that $X(t), X(s) \in V$.
Indeed, this follows immediately by noting that

$$\|M(t) - M(s)\|_H^2 - \|X(t) - X(s) - M(t) + M(s)\|_H^2$$
$$+ 2\langle X(s), M(t) - M(s)\rangle_H$$
$$= 2\langle X(t) - X(s), M(t) - M(s)\rangle_H - \|X(t) - X(s)\|_H^2$$
$$+ 2\langle X(s), M(t) - M(s)\rangle_H$$
$$= 2\langle X(t), M(t) - M(s)\rangle_H - \|X(t) - X(s)\|_H^2$$
$$= 2\langle X(t), X(t) - X(s)\rangle_H - 2 \int_s^t {}_{V^*}\langle Y(r), X(t)\rangle_V \, dr$$
$$- \|X(t)\|_H^2 - \|X(s)\|_H^2 + 2\langle X(t), X(s)\rangle_H$$
$$= \|X(t)\|_H^2 - \|X(s)\|_H^2 - 2 \int_s^t {}_{V^*}\langle Y(r), X(t)\rangle_V \, dr.$$

Claim (b): We have

$$E\left(\sup_{t\in[0,T]} \|X(t)\|_H^2\right) < \infty. \tag{4.2.9}$$

Indeed, by (4.2.8), for any $t = t_i^l \in I_l \setminus \{0, T\}$ given in Lemma 4.2.6,

$$\|X(t)\|_H^2 - \|X_0\|_H^2$$

$$= \sum_{j=0}^{i-1} (\|X(t_{j+1}^l)\|_H^2 - \|X(t_j^l)\|_H^2)$$

$$= 2\int_0^t {}_{V^*}\langle Y(s), \tilde{X}^l(s)\rangle_V \, ds$$

$$+ 2\int_0^t \langle \bar{X}^l(s), Z(s) \, dW(s)\rangle_H + 2\langle X(0), \int_0^{t_1^l} Z(s) \, dW(s)\rangle_H$$

$$+ \sum_{j=0}^{i-1} (\|M(t_{j+1}^l) - M(t_j^l)\|_H^2 - \|X(t_{j+1}^l) - X(t_j^l) - M(t_{j+1}^l) + M(t_j^l)\|_H^2). \tag{4.2.10}$$

We note that since $\bar{X}^l$ is pathwise bounded the stochastic integral involving $\bar{X}^l$ above is well-defined. By Lemma 4.2.6

$$E\left(\int_0^T |{}_{V^*}\langle Y(s), \tilde{X}^l(s)\rangle_V | \, ds\right) \leq \|Y\|_{K^*}\|\tilde{X}^l\|_K \leq c_1 \tag{4.2.11}$$

for some constant $c_1 > 0$ independent of l. Moreover, by the Burkholder–Davis inequality (cf. Proposition D.0.1), Lemmas 2.4.2 and 2.4.3,

$$E\left(\sup_{t\in[0,T]} \left| \int_0^t \langle \bar{X}^l(s), Z(s) \, dW(s)\rangle_H \right|\right)$$

$$\leqslant 3E\left(\left[\int_0^T \|Z(s)^*\bar{X}^l(s)\|_U^2 \, ds\right]^{1/2}\right)$$

$$\leqslant 3E\left(\left[\int_0^T \|\bar{X}^l(s)\|_H^2 \|Z(s)\|_{L_2(U,H)}^2 \, ds\right]^{1/2}\right) \tag{4.2.12}$$

$$= 3E\left(\left[\int_0^T \|\bar{X}^l(s)\|_H^2 \, d\langle M\rangle_s\right]^{1/2}\right)$$

$$\leq \frac{1}{4}E\left(\sup_{k_l-1\geq j\geq 0} \|X(t_j^l)\|_H^2\right) + 9E(\langle M\rangle_T),$$

where $\langle M \rangle_t = \int_0^t \|Z_{(s)}\|_{L_2(U,H)}^2 \, ds$ and we used that

$$ab \leqslant \frac{1}{12}a^2 + 3b^2, \ a, b > 0.$$

Finally, by Lemma 2.4.3

$$E\left(\sum_{j=0}^{i-1} \|M(t_{j+1}^l) - M(t_j^l)\|_H^2\right) = \sum_{j=0}^{i-1} E\left(\int_{t_j^l}^{t_{j+1}^l} \|Z(s)\|_{L_2(U,H)}^2 \, ds\right)$$

$$= E\left(\int_0^{t_i^l} \|Z(s)\|_{L_2(U,H)}^2 \, ds\right) \tag{4.2.13}$$

$$= E\left(\langle M \rangle_{t_i^l}\right).$$

Combining (4.2.10)–(4.2.13), we obtain

$$E\left(\sup_{t \in I_l \setminus \{T\}} \|X(t)\|_H^2\right) \leq c_2$$

for some constant $c_2 > 0$ independent of l. Therefore, letting $l \uparrow \infty$ and setting $I := \cup_{l \geq 1} I_l \setminus \{T\}$, with I_l as in Lemma 4.2.6, we obtain

$$E\left(\sup_{t \in I} \|X(t)\|_H^2\right) \leqslant c_2,$$

since $I_l \subset I_{l+1}$ for all $l \in \mathbb{N}$. Since for all $t \in [0, T]$

$$\sum_{j=1}^N {}_{V^*}\langle X(t), e_j \rangle_V^2 \uparrow \|X(t)\|_H^2 \text{ as } N \uparrow \infty,$$

where $\{e_j | j \in \mathbb{N}\} \subset V$ is an orthonormal basis of H and as usual for $x \in V^* \setminus H$ we set $\|x\|_H := \infty$, it follows that $t \mapsto \|X(t)\|_H$ is lower semicontinuous P-a.s. Since I is dense in $[0, T]$, we arrive at $\sup_{t \in [0,T]} \|X(t)\|_H^2 = \sup_{t \in I} \|X(t)\|_H^2$. Thus, (4.2.9) holds.
    Claim (c):

$$\lim_{l \to \infty} \sup_{t \in [0,T]} \left|\int_0^t \langle X(s) - \bar{X}^l(s), Z(s) \, dW(s)\rangle_H\right| = 0 \text{ in probability.} \tag{4.2.14}$$

We first note that because of (b) X is H-valued and by its continuity in V^* the process X is weakly continuous in H and, therefore, since $\mathcal{B}(H)$ is generated by H^*, progressively measurable as an H-valued process. Hence, for any $n \in \mathbb{N}$ the process $P_n X(s)$ is continuous in H so that

$$\lim_{l\to\infty} \int_0^T \|P_n(X(s) - \bar{X}^l(s))\|_H^2 \, \mathrm{d}\langle M\rangle_s = 0, \quad P\text{-a.s.}.$$

Here P_n denotes the orthogonal projection onto $\mathrm{span}\{e_1, \dots, e_n\}$ in H. Therefore, it suffices to show that for any $\varepsilon > 0$,

$$\lim_{n\to\infty} \sup_{l\in\mathbb{N}} P\left(\sup_{t\in[0,T]} \left| \int_0^t \langle (1 - P_n)\bar{X}^l(s), Z(s) \, \mathrm{d}W(s)\rangle_H \right| > \varepsilon \right) = 0,$$

$$\lim_{n\to\infty} P\left(\sup_{t\in[0,T]} \left| \int_0^t \langle (1 - P_n)X(s), Z(s) \, \mathrm{d}W(s)\rangle_H \right| > \varepsilon \right) = 0.$$

(4.2.15)

For any $n \in \mathbb{N}$, $\delta \in (0,1)$ and $N > 1$ by Corollary D.0.2 we have that

$$P\left(\sup_{t\in[0,T]} \left| \int_0^t \langle (1 - P_n)\bar{X}^l(s), Z(s) \, \mathrm{d}W(s)\rangle_H \right| > \varepsilon \right)$$

$$\leq \frac{3\delta}{\varepsilon} + P\left(\int_0^T \|\bar{X}^l(s)\|_H^2 \, \mathrm{d}\langle (1 - P_n)M\rangle_s > \delta^2 \right)$$

$$\leq \frac{3\delta}{\varepsilon} + P\left(\sup_{t\in[0,T]} \|X(t)\|_H > N \right) + \frac{N^2}{\delta^2} E\langle (1 - P_n)M\rangle_T.$$

By first letting $n \to \infty$, and using Lemma 2.4.3, and then letting $N \to \infty$ and finally $\delta \to 0$, we prove the first equality in (4.2.15). Similarly, the second equality is proved.

Claim (d): (4.2.4) holds for $t \in I$.

Fix $t \in I$. We may assume that $t \neq 0$. In this case for each sufficiently large $l \in \mathbb{N}$ there exists a unique $0 < i < k_l$ such that $t = t_i^l$. We have $X(t_j^l) \in V$ a.s. for all j. By Lemma 4.2.6 and (4.2.14) the sum of the first three terms in the right-hand side of (4.2.10) converges in probability to $2\int_0^t {}_{V^*}\langle Y(s), \bar{X}(s)\rangle_V \, \mathrm{d}s + 2\int_0^t \langle X(s), Z(s) \, \mathrm{d}W(s)\rangle_H$, as $l \to \infty$. Hence by Lemma 2.4.3

$$\|X(t)\|_H^2 - \|X(0)\|_H^2$$

$$= 2\int_0^t {}_{V^*}\langle Y(s), \bar{X}(s)\rangle_V \, \mathrm{d}s + 2\int_0^t \langle X(s), Z(s) \, \mathrm{d}W(s)\rangle_H + \langle M\rangle_t - \varepsilon_0,$$

where

$$\varepsilon_0 := P - \lim_{l\to\infty} \sum_{j=0}^{i-1} \|X(t_{j+1}^l) - X(t_j^l) - M(t_{j+1}^l) + M(t_j^l)\|_H^2$$

exists and "$P - \lim$" denotes limit in probability. So, to prove (4.2.4) for t as above, it suffices to show that $\varepsilon_0 = 0$. Since for any $\varphi \in V$,

$$\langle X(t_{j+1}^l) - X(t_j^l) - M(t_{j+1}^l) + M(t_j^l), \varphi \rangle_H = \int_{t_j^l}^{t_{j+1}^l} {}_{V^*}\langle Y(s), \varphi \rangle_V \, ds,$$

letting $\tilde{M}^l$ and $\bar{M}^l$ be defined as $\tilde{X}^l$ and $\bar{X}^l$ respectively, for M replacing X, we obtain for every $n \in \mathbb{N}$

$$\begin{aligned}
\varepsilon_0 = P - \lim_{l \to \infty} \Bigg(&\int_0^t {}_{V^*}\langle Y(s), \tilde{X}^l(s) - \bar{X}^l(s) - P_n(\tilde{M}^l(s) - \bar{M}^l(s)) \rangle_V \, ds \\
&- \langle X(t_1^l) - X(0) - M(t_1^l) + M(0), P_n M(0) - X(0) \rangle_H \\
&- \sum_{j=0}^{i-1} \langle X(t_{j+1}^l) - X(t_j^l) - M(t_{j+1}^l) \\
&+ M(t_j^l), (1 - P_n)(M(t_{j+1}^l) - M(t_j^l)) \rangle_H \Bigg).
\end{aligned}$$

By the weak continuity of X in H the second term converges to zero as $l \to \infty$. Lemma 4.2.6 implies that $\int_0^t {}_{V^*}\langle Y(s), \tilde{X}^l(s) - \bar{X}^l(s) \rangle_V \, ds \to 0$ in probability as $l \to \infty$. Moreover, since $P_n M(s)$ is a continuous process in V, $\int_0^t {}_{V^*}\langle Y(s), P_n(\tilde{M}^l(s) - \bar{M}^l(s)) \rangle_V \, ds \to 0$ as $l \to \infty$. Thus, by Lemma 2.4.3

$$\begin{aligned}
\varepsilon_0 \leq \text{P-} \lim_{l \to \infty} &\left(\sum_{j=0}^{i-1} \| X(t_{j+1}^l) - X(t_j^l) - M(t_{j+1}^l) + M(t_j^l) \|_H^2 \right)^{\frac{1}{2}} \\
&\cdot \left(\sum_{j=0}^{i-1} \| (1 - P_n)(M(t_{j+1}^l) - M(t_j^l)) \|_H^2 \right)^{\frac{1}{2}} \\
= &\, \varepsilon_0^{1/2} \langle (1 - P_n)M \rangle_t^{1/2},
\end{aligned}$$

which goes to zero as $n \to \infty$ again by Lemma 2.4.3 and Lebesgue's dominated convergence theorem. Therefore, $\varepsilon_0 = 0$.

Claim (e): (4.2.4) holds for all $t \in [0, T] \backslash I$.

Take $\Omega' \in \mathcal{F}$ with full probability such that the limit in (4.2.14) is a pointwise limit in Ω' for some subsequence (denoted again by $l \to \infty$) and (4.2.4) holds for all $t \in I$ on Ω'. If $t \notin I$, for any $l \in \mathbb{N}$ there exists a unique $j(l) < k_l$ such that $t \in]t_{j(l)}^l, t_{j(l)+1}^l]$. Letting $t(l) := t_{j(l)}^l$, we have $t(l) \uparrow t$ as $l \uparrow \infty$. By (4.2.4) for $t \in I$, for any $l > m$ we have on Ω' (since the above applies to

$X - X(t(m))$ replacing X)

$\|X(t(l)) - X(t(m))\|_H^2$

$$= 2 \int_{t(m)}^{t(l)} {}_{V^*}\langle Y(s), \bar{X}(s) - X(t(m)) \rangle_V \ ds$$

$$+ 2 \int_{t(m)}^{t(l)} \langle X(s) - X(t(m)), Z(s) \ dW(s) \rangle_H + \langle M \rangle_{t(l)} - \langle M \rangle_{(t(m)}$$

$$= 2 \int_0^T \mathbb{1}_{[t(m),t(l)]}(s) \, {}_{V^*}\langle Y(s), \bar{X}(s) - \bar{X}^m(s) \rangle_V \ ds$$

$$+ 2 \int_{t(l)}^{t(m)} \langle X(s) - \bar{X}^m(s), Z(s) \ dW(s) \rangle_H + \langle M \rangle_{t(l)} - \langle M \rangle_{(t(m)}.$$

$$(4.2.16)$$

The second summand is dominated by

$$4 \sup_{t \in [0,T]} \left| \int_0^t \langle X(s) - \bar{X}^m(s), Z(s) \ dW(s) \rangle_H \right|.$$

Thus, by the continuity of $\langle M \rangle_t$ and (4.2.14) (holding pointwise on Ω'), we have that

$$\lim_{m \to \infty} \sup_{l > m} \left\{ 2 \left| \int_0^T \mathbb{1}_{[t(m),t(l)]}(s) \langle \, X(s) - \bar{X}^m(s), Z(s) \ dW(s) \rangle_H \right| \right.$$

$$\left. + |\langle M \rangle_{t(l)} - \langle M \rangle_{(t(m)}| \right\} = 0$$

$$(4.2.17)$$

holding on Ω'. Furthermore, by Lemma 4.2.6, selecting another subsequence if necessary, we have for some $\Omega'' \in \mathcal{F}$ with full probability and $\Omega'' \subset \Omega'$, that on Ω''

$$\lim_{m \to \infty} \int_0^T |\,{}_{V^*}\langle Y(s), \bar{X}(s) - \bar{X}^m(s) \rangle_V | \ ds = 0.$$

Since for all $t \notin I$

$$\sup_{l > m} \int_{t(m)}^{t(l)} |\,{}_{V^*}\langle Y(s), \bar{X}(s) - \bar{X}^m(s) \rangle_V | \ ds$$

$$\leq \int_0^T |\,{}_{V^*}\langle Y(s), \bar{X}(s) - \bar{X}^m(s) \rangle_V | \ ds,$$

we have that

$$\lim_{m \to \infty} \sup_{l > m} \int_{t(m)}^{t(l)} {}_{V^*}\langle Y(s), \bar{X}(s) - \bar{X}^m(s) \rangle_V \, ds = 0$$

holds on Ω''.

Combining this with (4.2.16) and (4.2.17), we conclude that

$$\lim_{m \to \infty} \sup_{l \geq m} \| X(t(l)) - X(t(m)) \|_H^2 = 0$$

holds on Ω''. Thus, $(X(t(l)))_{l \in \mathbb{N}}$ converges in H on Ω''. Since we know that $X(t(l)) \to X(t)$ in V^*, it converges to $X(t)$ strongly in H on Ω''. Therefore, since (4.2.4) holds on Ω'' for $t(l)$, letting $l \to \infty$, we obtain (4.2.4) on Ω'' also for all $t \notin I$.

Claim (f): X is strongly continuous in H.

Since the right-hand side of (4.2.4) is on Ω'' continuous in $t \in [0, T]$, so must be its left-hand side, i.e. $t \mapsto \| X(t) \|_H$ is continuous on $[0, T]$. Therefore, the weak continuity of $X(t)$ in H implies its strong continuity in H.   □

Remark 4.2.8. *In the situation of Theorem 4.2.5 we have*

$$E(\| X(t) \|_H^2)$$

$$= E(\| X_0 \|_H^2) + \int_0^t E(2 \, {}_{V^*}\langle Y(s), \bar{X}(s) \rangle_V + \| Z(s) \|_{L_2(U,H)}^2) \, ds, \quad t \in [0, T].$$
$$(4.2.18)$$

Proof. Let $M(t)$, $t \in [0, T]$, denote the real valued local martingale in (4.2.4) and let τ_l, $l \in \mathbb{N}$, be $(\mathcal{F}_t)$-stopping times such that $M(t \wedge \tau_l)$, $t \in [0, T]$, is a martingale and $\tau_l \uparrow \infty$ as $l \to \infty$. Then for all $l \in \mathbb{N}$, $t \in [0, T]$, we have

$$E(\| X(t \wedge \tau_l) \|_H^2)$$

$$= E(\| X_0 \|_H^2) + \int_0^t E(1_{[0,\tau_l]}(s)[2 \, {}_{V^*}\langle Y(s), \bar{X}(s) \rangle_V + \| Z(s) \|_{L_2(U,H)}^2]) \, ds.$$
$$(4.2.19)$$

Using Claim (b) from the proof of Theorem 4.2.5 and the fact that the integrands on the right-hand side of (4.2.19) are $dt \otimes P$-integrable we can apply Lebesgue's dominated convergence theorem to obtain the assertion.   □

Now we turn to the proof of Theorem 4.2.4. We first need some preparations. Let $\{ e_i | i \in \mathbb{N} \} \subset V$ be an orthonormal basis of H and let $H_n := \text{span}\{ e_1, \dots, e_n \}$ such that $\text{span}\{ e_i | i \in \mathbb{N} \}$ is dense in V. Let $P_n : V^* \to H_n$ be defined by

$$P_n y := \sum_{i=1}^{n} {}_{V^*}\langle y, e_i \rangle_V e_i, \quad y \in V^*.$$
$$(4.2.20)$$

Clearly, $P_n|_H$ is just the orthogonal projection onto H_n in H. Let $\{g_i|i \in \mathbb{N}\}$ be an orthonormal basis of U and set

$$W^{(n)}(t) := \sum_{i=1}^{n} \langle W(t), g_i \rangle_U g_i = \sum_{i=1}^{n} B^i(t) g_i.$$

For each finite $n \in \mathbb{N}$ we consider the following stochastic equation on H_n:

$$\begin{aligned} \mathrm{d}X^{(n)}(t) \\ = P_n A(t, X^{(n)}(t)) \, \mathrm{d}t + P_n B(t, X^{(n)}(t)) \, \mathrm{d}W^{(n)}(t), \quad 1 \leq j \leq n, \end{aligned} \tag{4.2.21}$$

where $X^{(n)}(0) := P_n X_0$. It is easily seen (cf. in particular Remark 4.1.1, parts 1 and 2) that we are in the situation of Theorem 3.1.1 which implies that (4.2.21) has a unique continuous strong solution. Let

$$J := L^2([0,T] \times \Omega, \, \mathrm{d}t \otimes P; L_2(U,H)). \tag{4.2.22}$$

To construct the solution to (4.2.1), we need the following lemma.

Lemma 4.2.9. *Under the assumptions in Theorem 4.2.4, there exists $C \in \,]0, \infty[$ such that*

$$\|X^{(n)}\|_K + \|A(\cdot, X^{(n)})\|_{K^*} + \sup_{t \in [0,T]} E\|X^{(n)}(t)\|_H^2 \leq C \tag{4.2.23}$$

for all $n \in \mathbb{N}$.

Proof. By the finite-dimensional Itô formula we have P-a.s.

$$\begin{aligned} \|X^{(n)}(t)\|_H^2 = \|X_0^{(n)}\|_H^2 + \int_0^t \Big(2\,{}_{V^*}\langle A(s, X^{(n)}(s)), X^{(n)}(s) \rangle_V \\ + \|Z^{(n)}(s)\|_{L_2(U,H)}^2 \Big) \, \mathrm{d}s + M^{(n)}(t), \ t \in [0,T], \end{aligned}$$

where $Z^{(n)}(s) := P_n B(s, X^{(n)}(s))$ and

$$M^{(n)}(t) := 2\int_0^t \langle X^{(n)}(s), P_n B(s, X^{(n)}(s)) \, \mathrm{d}W^{(n)}(s) \rangle_H, \ t \in [0,T],$$

is a local martingale. Let τ_l, $l \in \mathbb{N}$, be $(\mathcal{F}_t)$-stopping times such that $\|X^{(n)}(t \wedge \tau_l)(\omega)\|_V$ is bounded uniformly in $(t,\omega) \in [0,T] \times \Omega$, $M^{(n)}(t \wedge \tau_l)$, $t \in [0,T]$, is a martingale for each $l \in \mathbb{N}$ and $\tau_l \uparrow \infty$ as $l \to \infty$. Then for all $l \in \mathbb{N}$, $t \in [0,T]$

$$\begin{aligned} E\left(\|X^{(n)}(t \wedge \tau_l)\|_H^2 \right) \\ = E(\|X_0^{(n)}\|_H^2) + \int_0^t E\Big(1_{[0,\tau_l]}(s)(2\,{}_{V^*}\langle A(s, X^{(n)}(s)), X^{(n)}(s) \rangle_V \\ + \|Z^{(n)}(s)\|_{L_2(U,H)}^2 \big) \Big) \, \mathrm{d}s. \end{aligned}$$

Hence using the product rule we obtain

$$E(e^{-c_1 t}\|X^{(n)}(t \wedge \tau_l)\|_H^2)$$

$$= E(\|X_0^{(n)}\|_H^2) + \int_0^t E(\|X^{(n)}(s \wedge \tau_l)\|_H^2) \, d(e^{-c_1 s})$$

$$+ \int_0^t e^{-c_1 s} \, d(E(\|X^{(n)}(s \wedge \tau_l)\|_H^2))$$

$$\quad\quad (4.2.24)$$

$$= E(\|X_0^{(n)}\|_H^2) - \int_0^t c_1 E(\|X^{(n)}(s \wedge \tau_l)\|_H^2) e^{-c_1 s} \, ds$$

$$+ \int_0^t e^{-c_1 s} E\Big(1_{[0,\tau_l]}(s)(2 \, {}_{V^*}\langle A(s, X^{(n)}(s)), X^{(n)}(s)\rangle_V$$

$$+ \|Z^{(n)}(s)\|_{L_2(U,H)}^2)\Big) \, ds.$$

Applying (H3) we arrive at

$$E(e^{-c_1 t}\|X^{(n)}(t \wedge \tau_l)\|_H^2) + \int_0^t c_1 E(\|X^{(n)}(s \wedge \tau_l)\|_H^2) e^{-c_1 s} \, ds$$

$$+ c_2 \int_0^t E(1_{[0,\tau_l]}(s)\|X^{(n)}(s \wedge \tau_l)\|_V^\alpha) e^{-c_1 s} \, ds$$

$$\leqslant E(\|X_0^{(n)}\|_H^2) + \int_0^t c_1 E(\|X^{(n)}(s)\|_H^2) e^{-c_1 s} \, ds + \int_0^T E(|f(s)|) \, ds.$$

Now taking $l \to \infty$ and applying Fatou's lemma we get

$$E(e^{-c_1 t}\|X^{(n)}(t)\|_H^2) + c_2 E\left(\int_0^t \|X^{(n)}(s)\|_V^\alpha e^{-c_1 s} \, ds\right)$$

$$\leqslant E(\|X_0^{(n)}\|_H^2) + E\left(\int_0^T |f(s)| \, ds\right)$$

for all $t \in [0, T]$. Here we used that by Theorem 4.2.5 (applied to (4.2.21)) the substracted terms are finite. Since $\|X_0^{(n)}\|_H \leqslant \|X_0\|_H$, now the assertion follows for the first and third summand in (4.2.23). For the remaining summand the assertion then follows by (H4). $\qquad\square$

Proof of Theorem 4.2.4. By the reflexivity of K, Lemma 4.2.9 and Remark 4.1.1, part 1, we have, for a subsequence $n_k \to \infty$:

(i) $X^{(n_k)} \to \bar{X}$ weakly in K and weakly in $L^2([0,T] \times \Omega;\ dt \otimes P; H)$.

(ii) $Y^{(n_k)} := A(\cdot, X^{(n_k)}) \to Y$ weakly in K^*.

(iii) $Z^{(n_k)} := P_{n_k} B(\cdot, X^{(n_k)}) \to Z$ weakly in J and hence

$$\int_0^{\cdot} P_{n_k} B(s, X^{(n_k)}(s))\, \mathrm{d}W^{(n_k)}(s) \to \int_0^{\cdot} Z(s)\, \mathrm{d}W(s)$$

weakly in $L^\infty([0,T],\ \mathrm{d}t; L^2(\Omega, P; H))$ (equipped with the supremum norm).

Here the second part in (iii) follows since also $B(\cdot, X^{(n_k)})\tilde{P}_{n_k} \to Z$ weakly in J, where $\tilde{P}_n$ is the orthogonal projection onto $\operatorname{span}\{g_1, \cdots, g_n\}$ in U, since

$$\int_0^{\cdot} P_{n_k} B(s, X^{(n_k)}(s))\, \mathrm{d}W^{(n_k)}(s) = \int_0^{\cdot} P_{n_k} B(s, X^{n_k}(s))\tilde{P}_{n_k}\, \mathrm{d}W(s)$$

and since a bounded linear operator between two Banach spaces is trivially weakly continuous. Since the approximants are progressively measurable, so are (the $\mathrm{d}t \otimes P$-versions) $\bar{X}, Y$ and Z.

Thus from (4.2.21) for all $v \in \bigcup_{n \geq 1} H_n$, $\varphi \in L^\infty([0,T] \times \Omega)$ by Fubini's theorem we get

$$E\left(\int_0^T {}_{V^*}\langle \bar{X}(t), \varphi(t)v\rangle_V\, \mathrm{d}t\right)$$

$$= \lim_{k\to\infty} E\left(\int_0^T {}_{V^*}\langle X^{(n_k)}(t), \varphi(t)v\rangle_V\, \mathrm{d}t\right)$$

$$= \lim_{k\to\infty} E\bigg(\int_0^T {}_{V^*}\langle X_0^{(n_k)}, \varphi(t)v\rangle_V\, \mathrm{d}t$$
$$+ \int_0^T \int_0^t {}_{V^*}\langle P_{n_k}Y^{(n_k)}(s), \varphi(t)v\rangle_V\, \mathrm{d}s\, \mathrm{d}t$$
$$+ \int_0^T \left\langle \int_0^t Z^{(n_k)}(s)\, \mathrm{d}W^{(n_k)}(s), \varphi(t)v\right\rangle_H \mathrm{d}t\bigg)$$

$$= \lim_{k\to\infty} \bigg[E\left(\langle X_0^{(n_k)}, v\rangle_H \int_0^T \varphi(t)\, \mathrm{d}t\right)$$
$$+ E\left(\int_0^T {}_{V^*}\langle Y^{(n_k)}(s), \int_s^T \varphi(t)\, \mathrm{d}t\, v\rangle_V\, \mathrm{d}s\right)$$
$$+ \int_0^T E\left(\varphi(t)\left\langle \int_0^t Z^{(n_k)}(s)\, \mathrm{d}W^{(n_k)}(s), v\right\rangle_H\right) \mathrm{d}t\bigg]$$

$$= E\left(\int_0^T {}_{V^*}\langle X_0 + \int_0^t Y(s)\, \mathrm{d}s + \int_0^t Z(s)\, \mathrm{d}W(s), \varphi(t)v\rangle_V\, \mathrm{d}t\right).$$

Therefore, defining

$$X(t) := X_0 + \int_0^t Y(s)\, ds + \int_0^t Z(s)\, dW(s), \quad t \in [0, T], \tag{4.2.25}$$

we have $X = \bar{X}$ $dt \otimes P$-a.e.

Now Theorem 4.2.5 applies to X in (4.2.25), so X is continuous in H and

$$E\left(\sup_{t \leq T} \|X(t)\|_H^2\right) < \infty.$$

Thus, it remains to verify that

$$B(\cdot, \bar{X}) = Z, \quad A(\cdot, \bar{X}) = Y, \quad dt \otimes P\text{-a.e.}. \tag{4.2.26}$$

To this end, we first note that for any nonnegative $\psi \in L^\infty([0, T], dt; \mathbb{R})$ it follows from (i) that

$$E\left(\int_0^T \psi(t)\|\bar{X}(t)\|_H^2\, dt\right)$$

$$= \lim_{k \to \infty} E\left(\int_0^T \langle \psi(t)\bar{X}(t), X^{(n_k)}(t)\rangle_H\, dt\right)$$

$$\leq \left(E\int_0^T \psi(t)\|\bar{X}(t)\|_H^2\, dt\right)^{1/2} \liminf_{k \to \infty} \left(E\int_0^T \psi(t)\|X^{(n_k)}(t)\|_H^2\, dt\right)^{1/2} < \infty.$$

Since $X = \bar{X}$ $dt \otimes P$-a.e., this implies

$$E\left(\int_0^T \psi(t)\|X(t)\|_H^2\, dt\right) \leq \liminf_{k \to \infty} E\left(\int_0^T \psi(t)\|X^{(n_k)}(t)\|_H^2\, dt\right). \tag{4.2.27}$$

By (4.2.25) using Remark 4.2.8 and the product rule we obtain that

$$E\left(e^{-ct}\|X(t)\|_H^2\right) - E\left(\|X_0\|_H^2\right)$$

$$= E\left(\int_0^t e^{-cs}\left(2\ _{V^*}\langle Y(s), \bar{X}(s)\rangle_V + \|Z(s)\|_{L_2(U,H)}^2 - c\|X(s)\|_H^2\right) ds\right). \tag{4.2.28}$$

Furthermore, for any $\phi \in K \cap L^2([0, T] \times \Omega, dt \otimes P; H)$ and taking $l \to \infty$ in

(4.2.24) with c_1 replaced by c

$$E\left(e^{-ct}\|X^{(n_k)}(t)\|_H^2\right) - E\left(\|X_0^{(n_k)}\|_H^2\right)$$

$$= E\left(\int_0^t e^{-cs}\left(2\,_{V^*}\langle A(s, X^{(n_k)}(s)), X^{(n_k)}(s)\rangle_V\right.\right.$$

$$\left.\left.+ \|P_{n_k}B(s, X^{(n_k)}(s))\tilde{P}_{n_k}\|_{L_2(U,H)}^2 - c\|X^{(n_k)}(s)\|_H^2\right)\,ds\right)$$

$$\leq E\left(\int_0^t e^{-cs}\left(2\,_{V^*}\langle A(s, X^{(n_k)}(s)), X^{(n_k)}(s)\rangle_V\right.\right.$$

$$\left.\left.+ \|B(s, X^{(n_k)}(s))\|_{L_2(U,H)}^2 - c\|X^{(n_k)}(s)\|_H^2\right)\,ds\right)$$

$$= E\left(\int_0^t e^{-cs}\left(2\,_{V^*}\langle A(s, X^{(n_k)}(s)) - A(s, \phi(s)), X^{(n_k)}(s) - \phi(s)\rangle_V\right.\right.$$

$$\left.\left.+ \|B(s, X^{(n_k)}(s)) - B(s, \phi(s))\|_{L_2(U,H)}^2 - c\|X^{(n_k)}(s) - \phi(s)\|_H^2\right)\,ds\right)$$

$$+ E\left(\int_0^t e^{-cs}\left(2\,_{V^*}\langle A(s, \phi(s)), X^{(n_k)}(s)\rangle_V\right.\right.$$

$$+ 2\,_{V^*}\langle A(s, X^{(n_k)}(s)) - A(s, \phi(s)), \phi(s)\rangle_V$$

$$- \|B(s, \phi(s))\|_{L_2(U,H)}^2 + 2\langle B(s, X^{(n_k)}(s)), B(s, \phi(s))\rangle_{L_2(U,H)}$$

$$\left.\left.- 2c\langle X^{(n_k)}(s), \phi(s)\rangle_H + c\|\phi(s)\|_H^2\right)\,ds\right).$$

$$(4.2.29)$$

Note that by (H2) the first of the two summands above is negative. Hence by letting $k \to \infty$ we conclude by (i)–(iii), Fubini's theorem, and (4.2.27) that for every nonnegative $\psi \in L^\infty([0, T],\, dt; \mathbb{R})$

$$E\left(\int_0^T \psi(t)(e^{-ct}\|X(t)\|_H^2 - \|X_0\|_H^2)\,dt\right)$$

$$\leq E\left(\int_0^T \psi(t)\left(\int_0^t e^{-cs}\Big[2\,_{V^*}\langle A(s, \phi(s)), \bar{X}(s)\rangle_V + 2\,_{V^*}\langle Y(s)\right.\right.$$

$$- A(s, \phi(s)), \phi(s)\rangle_V - \|B(s, \phi(s))\|_{L_2(U,H)}^2 + 2\langle Z(s), B(s, \phi(s))\rangle_{L_2(U,H)}$$

$$\left.\left.- 2c\langle X(s), \phi(s)\rangle_H + c\|\phi(s)\|_H^2\Big]\,ds\right)\,dt\right).$$

Inserting (4.2.28) for the left-hand side and rearranging as above we arrive at

$$0 \geq E\left(\int_0^T \psi(t) \left(\int_0^t e^{-cs} \left[2 \,_{V^*}\langle Y(s) - A(s, \phi(s)), \bar{X}(s) - \phi(s) \rangle_V \right. \right. \right.$$
$$\left. \left. \left. + \|B(s, \phi(s)) - Z(s)\|^2_{L_2(U,H)} - c\|X(s) - \phi(s)\|^2_H \right] ds \right) dt \right). \qquad (4.2.30)$$

Taking $\phi = \bar{X}$ we obtain from (4.2.30) that $Z = B(\cdot, \bar{X})$. Finally, first applying (4.2.30) to $\phi = \bar{X} - \varepsilon \tilde{\phi} v$ for $\varepsilon > 0$ and $\tilde{\phi} \in L^\infty([0,T] \times \Omega, \, dt \otimes P; \mathbb{R}), v \in V$, then dividing both sides by ε and letting $\varepsilon \to 0$, by Lebesgue's dominated convergence theorem, (H1) and (H4), we obtain

$$0 \geq E\left(\int_0^T \psi(t) \left(\int_0^t e^{-cs} \tilde{\phi}(s) \,_{V^*}\langle Y(s) - A(s, \bar{X}(s)), v \rangle_V \, ds \right) dt \right).$$

By the arbitrariness of ψ and $\tilde{\phi}$, we conclude that $Y = A(\cdot, \bar{X})$. This completes the existence proof.

The uniqueness is a consequence of the following proposition.   □

Proposition 4.2.10. *Consider the situation of Theorem 4.2.4 and let X, Y be two solutions. Then for $c \in \mathbb{R}$ as in (H2)*

$$E(\|X(t) - Y(t)\|^2_H) \leqslant e^{ct} E(\|X(0) - Y(0)\|^2_H) \text{ for all } t \in [0,T]. \qquad (4.2.31)$$

Proof. We first note that by our definition of solution (cf. Definition 4.2.1) and by Remark 4.1.1, part 1 we can apply Remark 4.2.8 to $X - Y$ and obtain for $t \in [0,T]$

$$E(\|X(t) - Y(t)\|^2_H) = E(\|X_0 - Y_0\|^2_H)$$
$$+ \int_0^t E(2 \,_{V^*}\langle A(s, \bar{X}(s)) - A(s, \bar{Y}(s)), \bar{X}(s) - \bar{Y}(s) \rangle_V$$
$$+ \|B(s, X(s)) - B(s, Y(s))\|^2_{L_2(U,H)}) \, ds$$
$$\leqslant E(\|X_0 - Y_0\|^2_H) + c \int_0^t E(\|X(s) - Y(s)\|^2_H) \, ds,$$

where we used (H2) in the last step. Applying Gronwall's lemma we obtain the assertion.   □

Remark 4.2.11. *Let $s \in [0,T]$ and $X_s \in L^2(\Omega, \mathcal{F}_s, P; H)$. Consider the equation*

$$X(t) = X_s + \int_s^t A(u, \bar{X}(u)) \, du + \int_s^t B(u, \bar{X}(u)) \, dW(u), \quad t \in [s,T] \quad (4.2.32)$$

with underlying Wiener process $W(t) - W(s)$, $t \in [s, T]$, and filtration $(\mathcal{F}_t)_{t \geqslant s}$, i.e. we just start our time at s. We define the notion of solution for (4.2.32) analogously to Definition 4.2.1. Then all results above in the case $s = 0$ carry over to this more general case. In particular, there exists a unique solution with initial condition X_s denoted by $X(t, s, X_s)$, $t \in [s, T]$. Let $0 \leqslant r \leqslant s \leqslant T$. Then for $X_r \in L^2(\Omega, \mathcal{F}_r, P; H)$

$$X(t, r, X_r) = X(t, s, X(s, r, X_r)), \quad t \in [s, T] \quad P\text{-}a.e. \tag{4.2.33}$$

Indeed, we have

$$X(t, r, X_r) = X_r + \int_r^t A(u, \bar{X}(u, r, X_r))\, du + \int_r^t B(u, \bar{X}(u, r, X_r))\, dW(u)$$

$$= X(s, r, X_r) + \int_s^t A(u, \bar{X}(u, r, X_r))\, du$$

$$+ \int_s^t B(u, \bar{X}(u, r, X_r))\, dW(u), \quad t \in [s, T].$$

But by definition $X(t, s, X(s, r, X_r))$, $t \in [s, T]$, satisfies the same equation. So, (4.2.33) follows by uniqueness. Furthermore, if for $s \in [0, T]$, $X_s = x$ for some $x \in H$ and A and B are independent of $\omega \in \Omega$, then $X(t, s, x)$ obviously is independent of $\mathcal{F}_s$ for all $t \in [s, T]$, since so are collections of increments of $W(t)$, $t \in [s, T]$.

4.3. Markov property and invariant measures

Now we are going to prove some qualitative results about the solutions of (4.2.1) or (4.2.32) and about their transition probabilities, i.e. about

$$p_{s,t}(x, dy) := P \circ (X(t, s, x))^{-1}(dy), \ 0 \leqslant s \leqslant t \leqslant T, \ x \in H. \tag{4.3.1}$$

As usual we set for $\mathcal{B}(H)$-measurable $F : H \to \mathbb{R}$, and $t \in [s, T]$, $x \in H$

$$p_{s,t}F(x) := \int F(y) p_{s,t}(x, dy),$$

provided F is $p_{s,t}(x, dy)$-integrable.

Remark 4.3.1. *The measures $p_{s,t}(x, dy)$, $0 \leqslant s \leqslant t \leqslant T$, $x \in H$, could in principle depend on the chosen Wiener process and the respective filtration. However, the construction of our solutions $X(t, s, x)$, $t \in [s, T]$, suggests that this is not the case. This can be rigorously proved in several ways. It is e.g. a consequence of the famous Yamada–Watanabe theorem which is included in Appendix E below in a finite-dimensional case, which immediately extends to infinite dimensions if the underlying Wiener process has covariance of finite trace. For the case of a cylindrical Wiener process we refer to [Ond04]. In these notes we shall use the latter as a fact referring to this remark each time we do so.*

Proposition 4.3.2. *Consider the situation of Theorem 4.2.4. Let $F : H \to \mathbb{R}$ be Lipschitz with*

$$Lip(F) := \sup_{x,y \in H, x \neq y} \frac{|F(x) - F(y)|}{\|x - y\|_H} (< \infty)$$

denoting its Lipschitz constant. Then for all $0 \leqslant s \leqslant t \leqslant T$

$$p_{s,t}|F|(x) < \infty \text{ for all } x \in H$$

and for all $x, y \in H$

$$|p_{s,t}F(x) - p_{s,t}F(y)| \leqslant e^{\frac{c}{2}(t-s)} Lip(F) \|x - y\|_H, \qquad (4.3.2)$$

where c is as in (H2).

Proof. Clearly, for all $x \in H$

$$|F(x)| \leqslant |F(0)| + Lip(F) \|x\|_H,$$

and thus for all $0 \leqslant s \leqslant t \leqslant T$

$$\begin{aligned}
p_{s,t}|F|(x) &= E(|F|(X(t,s,x))) \\
&\leqslant |F(0)| + Lip(F) E(\|X(t,s,x)\|_H) \\
&\leqslant |F(0)| + Lip(F) \left(E\left(\sup_{t \in [s,T]} \|X(t,s,x)\|_H^2 \right) \right)^{1/2} \\
&< \infty.
\end{aligned}$$

Furthermore, for $x, y \in H$ by (the "started at s" analogue of) (4.2.31)

$$\begin{aligned}
|p_{s,t}F(x) - p_{s,t}F(y)| &\leqslant E(|F(X(t,s,x)) - F(X(t,s,y)))|) \\
&\leqslant Lip(F) E(\|X(t,s,x) - X(t,s,y)\|_H) \\
&\leqslant Lip(F) e^{\frac{c}{2}(t-s)} \|x - y\|_H.
\end{aligned}$$

$\qquad\qquad\qquad\qquad\qquad\qquad\qquad\qquad\qquad\qquad\qquad\qquad\qquad\qquad\qquad \square$

Proposition 4.3.3. *Consider the situation of Theorem 4.2.4 and, in addition, assume that both A and B as well as f and g in (H3),(H4) respectively, are independent of $\omega \in \Omega$. Then any solution $X(t)$, $t \in [r,T]$, of (4.2.32) (with r replacing s) is Markov in the following sense:*
for every bounded, $\mathcal{B}(H)$-measurable $F : H \to \mathbb{R}$, and all $s, t \in [r,T]$, $s \leqslant t$

$$E(F(X(t))|\mathcal{F}_s)(\omega) = E(F(X(t,s,X(s)(\omega)))) \text{ for } P\text{-a.e. } \omega \in \Omega. \qquad (4.3.3)$$

Proof. Clearly, by a monotone class argument we may assume F in (4.3.3) to be Lipschitz continuous. We first note that by Proposition 4.3.2 for all $0 \leqslant s \leqslant t \leqslant T$ the map

$$H \ni x \mapsto E(F(X(t, s, x))) = p_{s,t} F(x)$$

is Lipschitz on H. So, the right-hand side of (4.3.3) is $\mathcal{F}_s$-measurable. Furthermore, for any bounded $\mathcal{F}_s$-measurable function $F_s : \Omega \to \mathbb{R}$, applying (4.2.33) we have

$$E(F_s F(X(t))) = E(F_s F(X(t, s, X(s)))). \tag{4.3.4}$$

By Lemma A.1.4 there exists a sequence of H-valued $\mathcal{F}_s$-measurable simple functions

$$f_n : \Omega \to H, \ f_n = \sum_{k=1}^{N_n} h_k^{(n)} 1_{\{f_n = h_k^{(n)}\}}, \quad N_n \in \mathbb{N},$$

where $h_1^{(n)}, \dots, h_{N_n}^{(n)} \in H$ are pairwise distinct and $\Omega = \bigcup_{k=1}^{N_n} \{f_n = h_k^{(n)}\}$, such that

$$\|f_n(\omega) - X(s)(\omega)\|_H \downarrow 0 \text{ as } n \to \infty \text{ for all } \omega \in \Omega.$$

Hence again by (the "s-shifted version" of) (4.2.31) the right-hand side of (4.3.4) is equal to

$$\lim_{n \to \infty} E(F_s F(X(t, s, f_n)))$$

$$= \lim_{n \to \infty} \sum_{k=1}^{N_n} E\left(F_s 1_{\{f_n = h_k^{(n)}\}} F(X(t, s, h_k^{(n)}))\right)$$

$$= \lim_{n \to \infty} \sum_{k=1}^{N_n} E\left(F_s 1_{\{f_n = h_k^{(n)}\}}\right) E\left(F(X(t, s, h_k^{(n)}))\right)$$

$$= \lim_{n \to \infty} E\left(F_s \sum_{k=1}^{N_n} 1_{\{f_n = h_k^{(n)}\}} E\left(F(X(t, s, h_k^{(n)}))\right)\right)$$

$$= \lim_{n \to \infty} \int F_s(\omega) E(F(X(t, s, f_n(\omega)))) P(d\omega)$$

$$= \int F_s(\omega) E(F(X(t, s, X(s)(\omega)))) P(d\omega),$$

where we used the last part of Remark 4.2.11 for the first and again (4.2.31) for the last equality. Now the assertion follows. $\qquad \square$

Corollary 4.3.4. *Consider the situation of Proposition 4.3.3 and let $0 \leqslant r \leqslant s \leqslant t \leqslant T$. Then*

$$p_{r,s} p_{s,t} = p_{r,t}, \tag{4.3.5}$$

i.e. for $F : H \to \mathbb{R}$, *bounded and* $\mathcal{B}(H)$-*measurable,* $x \in H$,

$$p_{r,s}(p_{s,t}F)(x) = p_{r,t}F(x).$$

Proof. For $F : H \to \mathbb{R}$ as above and $x \in H$ by Proposition 4.3.3 we have

$$p_{r,s}(p_{s,t}F)(x) = E(p_{s,t}F(X(s,r,x))) = \int E(F(X(t,s,X(s,r,x)(\omega))))P(d\omega)$$

$$= \int E(F(X(t,r,x))|\mathcal{F}_s)(\omega)P(d\omega)$$

$$= E(F(X(t,r,x))) = p_{r,t}F(x).$$

$\square$

Now let us assume that in the situation of Theorem 4.2.4 both A and B as well as f and g in (H3), (H4) respectively are independent of $(t, \omega) \in [0, T] \times \Omega$ (so they particularly hold for all $T \in [0, \infty[$). Then again using the notation introduced in Remark 4.2.11 for $0 \leqslant s \leqslant t < \infty$ and $x \in H$ we have

$$X(t, s, x) = X^{\tilde{W}}(t - s, 0, x) \text{ } P\text{-a.e.,} \tag{4.3.6}$$

where $X^{\tilde{W}}(t, 0, x)$, $t \in [0, \infty[$, is the solution of

$$X(t) = x + \int_0^t A(\bar{X}(u)) \, du + \int_0^t B(\bar{X}(u)) \, d\tilde{W}(u)$$

and $\tilde{W} := W(\cdot + s) - W(s)$ with filtration $\mathcal{F}_{s+u}$, $u \in [0, \infty[$, which is again a Wiener process. To show this let us express the dependence of the solution $X(t, s, x)$, $s \in [t, \infty)$ of (4.2.32) with $X_s := x$ on the Wiener process W by writing $X^W(t, s, x)$ instead of $X(t, s, x)$ and similarly, $p_{s,t}^W(s, dy)$ instead of $p_{s,t}(x, dy)$. Then, for all $0 \leqslant s \leqslant t < \infty$

$$X^W((t - s) + s, s, x)$$

$$= X^W(t, s, x)$$

$$= x + \int_s^t A(\bar{X}^W(u, s, x)) \, du + \int_s^t B(\bar{X}^W(u, s, x)) \, dW(u)$$

$$= x + \int_0^{t-s} A(\bar{X}^W(u + s, s, x)) \, du + \int_0^{t-s} B(\bar{X}^W(u + s, s, x)) \, d\tilde{W}(u),$$

So, by uniqueness the process $X^W(u+s, s, x)$, $u \in [0, \infty[$, must P-a.e. coincide with $X^{\tilde{W}}(u, 0, x)$, $u \in [0, \infty[$. In particular, it follows by Remark 4.3.1 that

$$p_{s,t}^W(x, dy) = P \circ (X^{\tilde{W}}(t - s, 0, x))^{-1}(dy) = p_{0,t-s}^{\tilde{W}}(x, dy) = p_{0,t-s}^W(x, dy) \tag{4.3.7}$$

("time homogeneity"), where we used Remark 4.3.1 for the last equality. Defining

$$p_t := p_{0,t}^W, \quad t \in [0, \infty[,$$

equality (4.3.5) for $r = 0$ and $s + t$ replacing t turns into

$$p_{s+t} = p_s p_t \quad \text{for } s, t \in [0, \infty[. \tag{4.3.8}$$

For $x \in H$ we define

$$P_x := P \circ (X(\cdot, 0, x))^{-1}, \tag{4.3.9}$$

i.e. P_x is the distribution of the solution to (4.2.1) with initial condition $x \in H$, defined as a measure on $C([0, \infty[, H)$. We equip $C([0, \infty[, H)$ with the σ-algebra

$$\mathcal{G} := \sigma(\pi_s | s \in [0, \infty[)$$

and filtration

$$\mathcal{G}_t := \sigma(\pi_s | s \in [0, t]), \, t \in [0, \infty[,$$

where $\pi_t(w) := w(t)$ for $w \in C([0, \infty[, H), \, t \in [0, \infty[.$

Proposition 4.3.5. *Consider the situation of Theorem 4.2.4 and, in addition, assume that both A and B as well as f and g in (H3),(H4) respectively, are independent of $(t, w) \in [0, T] \times \Omega$ (so they particularly hold for all $T \in [0, \infty[).$ Then the following assertions hold:*

1. *$P_x, x \in H$, form a time-homogenous Markov process on $C([0, \infty), H)$ with respect to the filtration $\mathcal{G}_t, t \in [0, \infty[$, i.e. for all $s, t \in [0, \infty[$, and all bounded, $\mathcal{B}(H)$-measurable $F : H \to \mathbb{R}$*

$$E_x(F(\pi_{t+s}) | \mathcal{G}_s) = E_{\pi_s}(F(\pi_t)) \quad P_x - a.e., \tag{4.3.10}$$

   where E_x and $E_x(\cdot | \mathcal{G}_s)$ denote expectation, conditional expectation with respect to P_x respectively.

2. *Suppose $\dim H < \infty$. If there exist $\eta, f \in]0, \infty[$ such that*

$$2 \, _{V^*}\langle A(v), v \rangle_V + \|B(v)\|_{L_2(U,H)}^2 \leqslant -\eta \|v\|_H^2 + f \quad \text{for all } v \in V, \tag{4.3.11}$$

   ("strict coercivity") then there exists an invariant measure μ for $(p_t)_{t \geqslant 0}$, i.e. μ is a probability measure on $(H, \mathcal{B}(H))$ such that

$$\int p_t F \, d\mu = \int F \, d\mu \quad \text{for all } t \in [0, \infty[\tag{4.3.12}$$

   and all bounded, $\mathcal{B}(H)$-measurable $F : H \to \mathbb{R}$.

Proof.   1. The right-hand side of (4.3.10) is $\mathcal{G}_s$-measurable by Proposition 4.3.2 and a monotone class argument. So, let $0 \leqslant t_1 < t_2 < \ldots < t_n \leqslant s$

and let $G : H^n \to \mathbb{R}$ be bounded and $\otimes_{i=1}^n \mathcal{B}(H)$-measurable. Then by (4.3.3) and (4.3.6)

$$E_x(G(\pi_{t_1}, \ldots, \pi_{t_n})F(\pi_{t+s}))$$
$$= E(G(X(t_1, 0, x), \ldots, X(t_n, 0, x))F(X(t+s, 0, x)))$$
$$= E(G(X(t_1, 0, x), \ldots, X(t_n, 0, x))E(F(X(t+s, 0, x))|\mathcal{F}_s))$$
$$= \int G(X(t_1, 0, x)(\omega), \ldots, X(t_n, 0, x)(\omega))$$
$$E(F(X(t+s, s, X(s, 0, x)(\omega))))P(d\omega)$$
$$= \int G(X(t_1, 0, x)(\omega), \ldots, X(t_n, 0, x)(\omega))$$
$$E(F(X(t, 0, X(s, 0, x)(\omega))))P(d\omega)$$
$$= \int G(\pi_{t_1}(\omega), \ldots, \pi_{t_n}(\omega))E(F(X(t, 0, \pi_s(\omega))))P_x(d\omega)$$
$$= \int G(\pi_{t_1}(\omega), \ldots, \pi_{t_n}(\omega))E_{\pi_s(\omega)}(F(\pi_t))P_x(d\omega).$$

Since the functions $G(\pi_{t_1}, \ldots, \pi_{t_n})$ considered above generate $\mathcal{F}_s$, equality (4.3.10) follows.

2. Let δ_0 be the Dirac measure in $0 \in H$ considered as a measure on $(H, \mathcal{B}(H))$ and for $n \in \mathbb{N}$ define the Krylov–Bogoliubov measure

$$\mu_n := \frac{1}{n} \int_0^n \delta_0 p_t \, dt,$$

i.e. for $\mathcal{B}(H)$-measurable $F : H \to [0, \infty[$

$$\int F \, d\mu_n = \frac{1}{n} \int_0^n p_t F(0) \, dt.$$

Clearly, each μ_n is a probability measure. We first prove that $\{\mu_n | n \in \mathbb{N}\}$ is tight. By Remark 4.2.8 for any solution X to (4.2.1) applying the product rule and using (4.3.11) we get that

$$E(e^{\eta t} \|X(t)\|_H^2) = E(\|X(0)\|_H^2) + E\left(\int_0^t e^{\eta s} \left(2 \,_{V^*}\langle A(\bar{X}(s)), \bar{X}(s)\rangle_V \right. \right.$$
$$+ \|B(\bar{X}(s))\|_{L_2(U,H)}^2 + \eta \|\bar{X}(s)\|_H^2 \bigg) \, ds \bigg)$$
$$\leq E(\|X(0)\|_H^2) + f \int_0^t e^{\eta s} \, ds, \quad t \in [0, \infty[.$$

Therefore,

$$E(\|X(t)\|_H^2) \leqslant e^{-\eta t} E(\|X(0)\|_H^2) + \frac{f}{\eta}, \quad t \in [0, \infty[, \qquad (4.3.13)$$

which in turn implies that

$$\int \|x\|_H^2 \mu_n(\mathrm{d}x) = \frac{1}{n} \int_0^n E(\|X(t,0,0)\|_H^2)\, \mathrm{d}t \leqslant \frac{f}{\eta} \quad \text{for all } n \in \mathbb{N}.$$
$$(4.3.14)$$

Hence by Chebychev's inequality

$$\sup_{n \in \mathbb{N}} \mu_n(\{\|\cdot\|_H^2 > R\}) \leqslant \frac{1}{R} \frac{f}{\eta} \to 0 \text{ as } R \to \infty. \qquad (4.3.15)$$

Since $\dim H < \infty$, the closed balls $\{\|\cdot\|_H^2 \leqslant R\}$, $R \in]0, \infty[$, are compact. Hence by Prohorov's theorem there exists a probability measure μ and a subsequence $(\mu_{n_k})_{k \in \mathbb{N}}$ such that $\mu_{n_k} \to \mu$ weakly as $k \to \infty$.

Now let us prove that μ is invariant for $(p_t)_{t \geqslant 0}$. So, let $t \in [0, \infty[$ and let $F : H \to \mathbb{R}$ be bounded and $\mathcal{B}(H)$-measurable. By a monotone class argument we may assume that F is Lipschitz continuous. Then $p_t F$ is bounded and (Lipschitz) continuous by Proposition 4.3.2. Hence using (4.3.8) for the third equality below, we obtain

$$\int p_t F \, \mathrm{d}\mu$$

$$= \lim_{k \to \infty} \int p_t F \, \mathrm{d}\mu_{n_k}$$

$$= \lim_{k \to \infty} \frac{1}{n_k} \int_0^{n_k} p_s(p_t F)(0) \, \mathrm{d}s$$

$$= \lim_{k \to \infty} \frac{1}{n_k} \int_0^{n_k} p_{s+t} F(0) \, \mathrm{d}s$$

$$= \lim_{k \to \infty} \int F \, \mathrm{d}\mu_{n_k} + \lim_{k \to \infty} \frac{1}{n_k} \int_{n_k}^{n_k+t} p_s F(0) \, \mathrm{d}s - \lim_{k \to \infty} \frac{1}{n_k} \int_0^t p_s F(0) \, \mathrm{d}s$$

$$= \int F \, \mathrm{d}\mu,$$

$$(4.3.16)$$

since $|p_s F(0)| \leqslant \sup_{x \in H} |F(x)|$, so the second and third limits above are equal to zero.

$\square$

Remark 4.3.6. If $\dim H = \infty$, the above proof of Proposition 4.3.5, part 2 works up to and including (4.3.15). However, since closed balls are no longer *compact, one can apply Prohorov's theorem only on a Hilbert space H_1 into which H is compactly embedded. So, let H_1 be a separable Hilbert space such that $H \subset H_1$ compactly and densely (e.g. take H_1 to be the completion of H in the norm*

$$\|x\|_1 := \left[\sum_{i=1}^{\infty} \alpha_i \langle x, e_i \rangle_H^2 \right]^{1/2}, \ x \in H,$$

where $\alpha_i \in]0, \infty[$, $\sum_{i=1}^{\infty} \alpha_i < \infty$, and $\{e_i | i \in \mathbb{N}\}$ is an orthonormal basis of H); extending the measures μ_n by zero to $\mathcal{B}(H_1)$ we obtain that $\{\mu_n | n \in \mathbb{N}\}$ is tight on H_1. This extension of the measures is possible, since by Kuratowski's theorem $H \in \mathcal{B}(H_1)$ and $\mathcal{B}(H_1) \cap H = \mathcal{B}(H)$. Hence by Prohorov's theorem there exists a probability measure $\bar{\mu}$ on $(H_1, \mathcal{B}(H_1))$ and a subsequence $(\mu_{n_k})_{k \in \mathbb{N}}$ such that $\mu_{n_k} \to \bar{\mu}$ weakly on H_1 as $k \to \infty$. As in Exercise 4.2.3, part 1 one constructs a lower semicontinuous function $\Theta : H_1 \to [0, \infty]$ such that

$$\Theta := \begin{cases} \|\cdot\|_H & on \ H \\ +\infty & on \ H_1 \backslash H. \end{cases}$$

Then (4.3.14) implies that for $l_i, i \in \mathbb{N}$, as in Example 4.2.3, part 1,

$$\int_{H_1} \Theta^2(x) \bar{\mu}(\, dx) = \lim_{N \to \infty} \lim_{M \to \infty} \int \sup_{i \leqslant N} \langle l_i, x \rangle_{H_1}^2 \wedge M \bar{\mu}(dx)$$

$$= \sup_{M, N \in \mathbb{N}} \lim_{k \to \infty} \int \sup_{i \leqslant N} \langle l_i, x \rangle_{H_1}^2 \wedge M \mu_{n_k}(dx)$$

$$\leqslant \liminf_{k \to \infty} \sup_{N, M \in \mathbb{N}} \int \sup_{i \leqslant N} \langle l_i, x \rangle_{H_1}^2 \wedge M \mu_{n_k}(dx)$$

$$= \liminf_{k \to \infty} \int_H \|x\|_H^2 \mu_{n_k}(dx)$$

$$\leqslant \frac{f}{\eta}.$$

Hence $\Theta < \infty$ $\bar{\mu}$-a.e., so $\bar{\mu}(H) = 1$. Therefore, $\mu := \bar{\mu}\big|_{\mathcal{B}(H)}$ is a probability measure on $(H, \mathcal{B}(H))$.

Unfortunately, the part of the proof of Proposition 4.3.5, part 2 above, which shows that μ is invariant, does not work. More precisely, for the first equality in (4.3.16) we need that $p_t F$ is continuous with respect to the same topology with respect to which $(\mu_{n_k})_{k \in \mathbb{N}}$ converges weakly, i.e. the topology on H_1. This one is, however, weaker than that on H. So, unless we can construct H_1 in such a way that $p_t F$ has a continuous extension to H_1, the first equality in (4.3.16) may not hold.

So far, we have taken a positive time s as the starting time for our SDE (see Remark 4.2.11). In the case of coefficients independent of t and ω, it is possible and convenient to consider negative starting times also. For this we, however, need a Wiener process with negative time. To this end we recall that we can run a cylindrical Wiener process $W(t), t \in [0, \infty[$ on H (with positive time) backwards in time and get again a Wiener process. More precisely, for fixed $T \in [0, \infty[$ we have that $W(T - t) - W(T)$, $t \in [0, T]$ is again a cylindrical Wiener process with respect to the filtration $\sigma(\{W(T - s) - W(T)|s \in [0, t]\})$, $t \in [0, T]$, and also with respect to the filtration $\sigma(\{W(r_2) - W(r_1)|r_1, r_2 \in [T - t, \infty[, r_2 \leqslant r_1\})$, $t \in [0, T]$, where the latter will be more convenient for us.

So, let A, B be independent of $(t, \omega) \in [0, T] \times \Omega$ and let $W^{(1)}(t)$, $t \in [0, \infty[$, be another cylindrical Wiener process on $(\Omega, \mathcal{F}, P)$ with covariance operator $Q = I$, independent of $W(t)$, $t \in [0, \infty[$. Define

$$\bar{W}(t) := \begin{cases} W(t), & \text{if } t \in [0, \infty[, \\ W^{(1)}(-t), & \text{if } t \in]-\infty, 0] \end{cases} \qquad (4.3.17)$$

with filtration

$$\bar{\mathcal{F}}_t := \bigcap_{s > t} \bar{\mathcal{F}}_s^\circ, \quad t \in \mathbb{R}, \qquad (4.3.18)$$

where $\bar{\mathcal{F}}_s^\circ := \sigma(\{\bar{W}(r_2) - \bar{W}(r_1)|r_1, r_2 \in]-\infty, s], r_2 \geqslant r_1\}, \mathcal{N})$ and $\mathcal{N} := \{A \in \mathcal{F}|P(A) = 0\}$. As in the proof of Proposition 2.1.13 one shows that if $-\infty < s < t < \infty$, then $\bar{W}(t) - \bar{W}(s)$ is independent of $\bar{\mathcal{F}}_s$. Now for $s \in \mathbb{R}$ fixed consider the SDE

$$\mathrm{d}X(t) = A(X(t))\,\mathrm{d}t + B(X(t))\,\mathrm{d}\bar{W}(t), \, t \in [s, \infty[. \qquad (4.3.19)$$

Remark 4.3.7. *Let $s \in \mathbb{R}$ and $X_s \in L^2(\Omega, \bar{\mathcal{F}}_s, P; H)$ and consider the integral version of (4.3.19)*

$$X(t) = X_s + \int_s^t A(\bar{X}(u))\,\mathrm{d}u + \int_s^t B(\bar{X}(u))\,\mathrm{d}\bar{W}(u), \quad t \in [s, \infty[, \quad (4.3.20)$$

with underlying Wiener process $\bar{W}(t) - \bar{W}(s)$, $t \in [s, \infty[$ and filtration $(\bar{\mathcal{F}}_t)_{t \geqslant s}$ (cf. Remark 4.2.11). We define the notion of solution for (4.3.20) analogously to Definition 4.2.1. Then again all results above for $s = 0$ (respectively for $s \in [0, \infty[$, see Remark 4.2.11) carry over to this more general case. In particular, we have the analogue of (4.3.5), namely

$$p_{r,s}p_{s,t} = p_{r,t} \quad \text{for all } -\infty < r \leqslant s \leqslant t < \infty, \qquad (4.3.21)$$

where for $s, t \in \mathbb{R}$, $s \leqslant t, x \in H$

$$p_{s,t}(x, dy) := P \circ (X(t, s, x))^{-1}(dy),$$

and analogously to (4.3.7) one shows that

$$p_{s,t}(x, dy) = p_{0,t-s}(x, dy).$$

In particular, for $t = 0$ we have

$$p_{-s,0}(x, dy) = p_{0,s}(x, dy) \quad \text{for all } x \in H, \ s \in [0, \infty[. \tag{4.3.22}$$

Furthermore, for every $s \in \mathbb{R}$ there exists a unique solution with initial condition X_s denoted by $X(t, s, X_s)$, $t \in [s, \infty[$, and (4.2.33) as well as the final part of Remark 4.2.11 hold also in this case.

Our next main aim (cf. Theorem 4.3.9 below) is to prove the existence of a unique invariant measure for (4.3.19) if the constant c in (H2) is strictly negative ("strict monotonicity"). The method of the proof is an adaptation from [DPZ96, Subsection 6.3.1]. We shall need the following:

Lemma 4.3.8. *Suppose (H3), (H4) hold and that (H2) holds for $c := -\lambda$ for some $\lambda \in]0, \infty[$. Let $\eta \in]0, \lambda[$. Then there exists $\delta_\eta \in]0, \infty[$ such that for all $v \in V$*

$$2 \,_{V^*}\langle A(v), v \rangle_V + \|B(v)\|_{L_2(U,H)}^2 \leqslant -\eta \|v\|_H^2 + \delta_\eta. \tag{4.3.23}$$

Proof. Let $v \in V$ and $\varepsilon \in]0, 1[$. Then using (H2) first (with $c = -\lambda$ according to our assumption), then Remark 4.1.1, part 1 and finally (H3) we obtain

$$2 \,_{V^*}\langle A(v), v \rangle_V + \|B(v)\|_{L_2(U,H)}^2$$

$$= 2 \,_{V^*}\langle A(v) - A(0), v \rangle_V + 2 \,_{V^*}\langle A(0), v \rangle_V + \|B(v) - B(0)\|_{L_2(U,H)}^2$$

$$- \|B(0)\|_{L_2(U,H)}^2 + 2\langle B(v), B(0) \rangle_{L_2(U,H)}$$

$$\leqslant -\lambda \|v\|_H^2 + 2\varepsilon \|v\|_V^\alpha + 2\varepsilon^{-\frac{1}{\alpha-1}}(\alpha-1)\alpha^{\frac{\alpha}{\alpha-1}}\|A(0)\|_{V^*}^{\frac{\alpha}{\alpha-1}} + \varepsilon^{-1}\|B(0)\|_{L_2(U,H)}^2$$

$$+ \varepsilon \|B(v)\|_{L_2(U,H)}^2$$

$$\leqslant -\lambda \|v\|_H^2 + 2\varepsilon \|v\|_V^\alpha + \beta_\varepsilon$$

$$+ \varepsilon \left(c_1 \|v\|_H^2 + f + \frac{2}{\alpha}\|v\|_V^\alpha + 2\frac{\alpha-1}{\alpha}g^{\frac{\alpha}{\alpha-1}} + 2c_3 \|v\|_V^\alpha \right)$$

$$\leqslant \left[-\lambda + \varepsilon c_1 \left(1 + \frac{2}{c_2}(1 + \alpha^{-1} + c_3) \right) \right] \|v\|_H^2 + \tilde{\beta}_\varepsilon + \frac{2}{c_2}\varepsilon(1 + \alpha^{-1} + c_3)f$$

$$- \frac{2}{c_2}\varepsilon(1 + \alpha^{-1} + c_3)(2 \,_{V^*}\langle A(v), v \rangle_V + \|B(v)\|_{L_2(U,H)}^2)$$

with $\beta_\varepsilon, \tilde{\beta}_\varepsilon \in]0, \infty[$ independent of v and where we applied Young's inequality in the form

$$ab = [(\alpha\varepsilon)^{-1/\alpha}a][(\alpha\varepsilon)^{1/\alpha}b] \leqslant \frac{(\alpha\varepsilon)^{-1/(\alpha-1)}}{\alpha/(\alpha-1)}a^{\alpha/(\alpha-1)} + \varepsilon b^\alpha,$$

$a, b \in [0, \infty[$ in the second step. Hence taking ε small enough we can find $\delta_\eta \in]0, \infty[$ such that for all $v \in V$

$$2 \,_{V^*}\langle A(v), v \rangle_V + \|B(v)\|_{L_2(U,H)}^2 \leqslant -\eta \|v\|_H^2 + \delta_\eta.$$

$\square$

Theorem 4.3.9. *Consider the situation of Proposition 4.3.5 and, in addition, assume that $c \in \mathbb{R}$ in (H2) is strictly negative, i.e. $c = -\lambda$, $\lambda \in]0, \infty[$ ("strict monotonicity"). Then there exists an invariant measure μ for $(p_t)_{t \geqslant 0}$ such that*

$$\int \|y\|_H^2 \mu(dy) < \infty.$$

Moreover, for $F : H \to \mathbb{R}$ Lipschitz, $x \in H$ and any invariant measure μ for $(p_t)_{t \geqslant 0}$

$$\left| p_t F(x) - \int F \, d\mu \right| \leqslant e^{-\frac{\lambda}{2}t} Lip(F) \int \|x - y\|_H \mu(dy) \quad \text{for all } t \in [0, \infty[.$$
$$(4.3.24)$$

In particular, there exists exactly one invariant measure for $(p_t)_{t \geqslant 0}$ with the property that

$$\int \|y\|_H \mu(dy) < \infty.$$

Remark 4.3.10. (4.3.24) *is referred to as "exponential convergence of $(p_t)_{t \geqslant 0}$ to equilibrium" (uniformly with respect to x in balls in H).*

For the proof of Theorem 4.3.9 we need one lemma.

Lemma 4.3.11. *Consider the situation of Theorem 4.3.9. Let $t \in \mathbb{R}$. Then there exists $\eta_t \in L^2(\Omega, \mathcal{F}, P; H)$, such that for all $x \in H$*

$$\lim_{s \to -\infty} X(t, s, x) = \eta_t \quad in \; L^2(\Omega, \mathcal{F}, P; H).$$

Moreover, there exists $C \in [0, \infty[$ such that for all $s \in] - \infty, t]$

$$E(\|X(t, s, x) - \eta_t\|_H^2) \leqslant C e^{\lambda(s-t)} (1 + \|x\|_H^2).$$

Proof. For $s_1, s_2 \in] - \infty, t]$, $s_1 \leqslant s_2$, and $x \in H$

$$X(t, s_1, x) - X(t, s_2, x)$$

$$= \int_{s_2}^t [A(\bar{X}(u, s_1, x)) - A(\bar{X}(u, s_2, x))] \, ds$$

$$\quad + \int_{s_2}^t [B(\bar{X}(u, s_1, x)) - B(\bar{X}(u, s_2, x))] \, d\bar{W}(u) + X(s_2, s_1, x) - x,$$

since

$$X(s_2, s_1, x) = x + \int_{s_1}^{s_2} A(\bar{X}(u, s_1, x)) \, du + \int_{s_1}^{s_2} B(\bar{X}(u, s_1, x)) \, d\bar{W}(u).$$
$$(4.3.25)$$

Since Remark 4.2.8 extends to our present case we can use the product rule and (H2) with $c = -\lambda$ to obtain

$$E(e^{\lambda t}\|X(t,s_1,x) - X(t,s_2,x)\|_H^2) = E(e^{\lambda s_2}\|X(s_2,s_1,x) - x\|_H^2)$$

$$+ \int_{s_2}^t e^{\lambda u} E\Big(2\,_{V^*}\langle A(\bar{X}(u,s_1,x)) - A(\bar{X}(u,s_2,x)), \bar{X}(u,s_1,x) - \bar{X}(u,s_1,x)\rangle_V$$

$$+ \|B(\bar{X}(u,s_1,x)) - B(\bar{X}(u,s_2,x))\|_{L_2(U,H)}^2\Big)\,du$$

$$+ \int_{s_2}^t e^{\lambda u}\lambda E\left(\|X(u,s_1,x) - X(u,s_2,x)\|_H^2\right)\,du$$

$$\leqslant 2e^{\lambda s_2}[E\left(\|X(s_2,s_1,x)\|_H^2\right) + \|x\|_H^2].$$

$$(4.3.26)$$

But again by Remark 4.2.8 extended to the present case, the product rule and (4.3.23) imply

$$E(e^{\eta s_2}\|X(s_2,s_1,x)\|_H^2)$$

$$= e^{s_1\eta}\|x\|_H^2 + \int_{s_1}^{s_2} e^{\eta u} E\Big(2\,_{V^*}\langle A(\bar{X}(u,s_1,x)), \bar{X}(u,s_1,x)\rangle_V$$

$$+ \|B(\bar{X}(u,s_1,x))\|_{L_2(U,H)}^2\Big)\,du + \int_{s_1}^{s_2} e^{\eta u}\eta E(\|X(u,s_1,x)\|_H^2)\,du \qquad (4.3.27)$$

$$\leqslant e^{s_1\eta}\|x\|_H^2 + \delta_\eta \int_{s_1}^{s_2} e^{\eta u}\,du \leqslant e^{s_1\eta}\|x\|_H^2 + \frac{\delta_\eta}{\eta}e^{s_2\eta}.$$

Combining (4.3.26) and (4.3.27) we obtain

$$E(\|X(t,s_1,x) - X(t,s_2,x)\|_H^2) \leqslant 2\left(\frac{\delta_\eta}{\eta} + 2\|x\|_H^2\right)e^{\lambda(s_2-t)}. \qquad (4.3.28)$$

Letting s_2 (hence s_1) tend to $-\infty$, it follows that there exists $\eta_t(x) \in L^2(\Omega, \mathcal{F}, P; H)$ such that

$$\lim_{s\to-\infty} X(t,s,x) = \eta_t(x) \text{ in } L^2(\Omega, \mathcal{F}, P; H),$$

and letting $s_1 \to -\infty$ in (4.3.28) the last part of the assertion follows also, provided we can prove that $\eta_t(x)$ is independent of $x \in H$. To this end let

$x, y \in H$ and $s \in]-\infty, t]$. Then

$$X(t, s, x) - X(t, s, y)$$

$$= x - y + \int_s^t (A(\bar{X}(u, s, x)) - A(\bar{X}(u, s, y))) \, du$$

$$+ \int_s^t (B(\bar{X}(u, s, x)) - B(\bar{X}(u, s, y))) \, d\bar{W}(u).$$

Hence by the same arguments to derive (4.3.26) we get

$$E(e^{\lambda t} \|X(t, s, x) - X(t, s, y)\|_H^2) \leqslant e^{\lambda s} \|x - y\|_H^2,$$

so

$$\lim_{s \to -\infty} (X(t, s, x) - X(t, s, y)) = 0 \text{ in } L^2(\Omega, \mathcal{F}, P; H).$$

Hence both assertions are completely proved. $\qquad\square$

Proof of Theorem 4.3.9. Define

$$\mu := P \circ \eta_0^{-1}$$

with η_0 as in Lemma 4.3.11. Since $\eta_0 \in L^2(\Omega, \mathcal{F}, P; H)$ we have that

$$\int \|y\|_H^2 \mu(dy) < \infty.$$

Let $t \in [0, \infty[$. We note that by (4.3.21) and (4.3.22) for all $s \in [0, \infty[$

$$p_{-s,0} p_{0,t} = p_{-s,t} = p_{0,t+s} = p_{-(t+s),0}. \qquad (4.3.29)$$

Let $F : H \to \mathbb{R}$, F bounded and Lipschitz. Then by Proposition 4.3.2 we have that $p_{0,t} F$ is (bounded and) Lipschitz. Furthermore, by Lemma 4.3.11 for all $x \in H$

$$p_{-s,0}(x, dy) \to \mu \text{ weakly as } s \to \infty.$$

Hence by (4.3.29) for all $x \in H$

$$\int p_{0,t} F \, d\mu = \lim_{s \to \infty} p_{-s,0}(p_{0,t} F)(x) = \lim_{s \to \infty} p_{-(t+s),0} F(x) = \int F \, d\mu.$$

Recalling that by definition $p_t = p_{0,t}$, it follows that μ is an invariant measure for $(p_t)_{t \geqslant 0}$. Furthermore, if μ is an invariant measure for $(p_t)_{t \geqslant 0}$, then by Proposition 4.3.2 for all $t \in [0, \infty[$

$$\left| p_t F(x) - \int F \, d\mu \right| = \left| \int (p_t F(x) - p_t F(y)) \mu(dy) \right|$$

$$\leqslant e^{-\frac{\lambda}{2} t} \text{Lip}(F) \int \|x - y\|_H \mu(dy).$$

$\qquad\square$

A. The Bochner Integral

This chapter is a slight modification of Chap. A in [FK01].

Let $(X, \| \ \|)$ be a Banach space, $\mathcal{B}(X)$ the Borel σ-field of X and $(\Omega, \mathcal{F}, \mu)$ a measure space with finite measure μ.

A.1. Definition of the Bochner integral

Step 1: As first step we want to define the integral for simple functions which are defined as follows. Set

$$\mathcal{E} := \left\{ f : \Omega \to X \ \Big| \ f = \sum_{k=1}^{n} x_k 1_{A_k}, \ x_k \in X, \ A_k \in \mathcal{F}, \ 1 \leqslant k \leqslant n, \ n \in \mathbb{N} \right\}$$

and define a semi-norm $\| \ \|_{\mathcal{E}}$ on the vector space $\mathcal{E}$ by

$$\|f\|_{\mathcal{E}} := \int \|f\| \, \mathrm{d}\mu, \quad f \in \mathcal{E}.$$

To get that $(\mathcal{E}, \| \ \|_{\mathcal{E}})$ is a normed vector space we consider equivalence classes with respect to $\| \ \|_{\mathcal{E}}$. For simplicity we will not change the notations.

For $f \in \mathcal{E}$, $f = \sum_{k=1}^{n} x_k 1_{A_k}$, A_k's *pairwise disjoint* (such a representation is called *normal* and always exists, because $f = \sum_{k=1}^{n} x_k 1_{A_k}$, where $f(\Omega) = \{x_1, \ldots, x_k\}$, $x_i \neq x_j$, and $A_k := \{f = x_k\}$) and we now define the Bochner integral to be

$$\int f \, \mathrm{d}\mu := \sum_{k=1}^{n} x_k \mu(A_k).$$

(Exercise: This definition is independent of representations, and *hence* linear.) In this way we get a mapping

$$\mathrm{int} : \ (\mathcal{E}, \| \ \|_{\mathcal{E}}) \ \to \ (X, \| \ \|)$$
$$f \ \mapsto \ \int f \, \mathrm{d}\mu$$

which is linear and uniformly continuous since $\left\| \int f \, \mathrm{d}\mu \right\| \leqslant \int \|f\| \, \mathrm{d}\mu$ for all $f \in \mathcal{E}$.

Therefore we can extend the mapping int to the abstract completion of $\mathcal{E}$ with respect to $\| \ \|_{\mathcal{E}}$ which we denote by $\overline{\mathcal{E}}$.

Step 2:   We give an explicit representation of $\overline{\mathcal{E}}$.

Definition A.1.1. A function $f : \Omega \to X$ is called strongly measurable if it is $\mathcal{F}/\mathcal{B}(X)$-measurable and $f(\Omega) \subset X$ is separable.

Definition A.1.2. Let $1 \leqslant p < \infty$. Then we define

$$\mathcal{L}^p(\Omega, \mathcal{F}, \mu; X) := \mathcal{L}^p(\mu; X)$$
$$:= \left\{ f : \Omega \to X \,\middle|\, f \text{ is strongly measurable with} \right.$$
$$\left. \text{respect to } \mathcal{F}, \text{ and } \int \|f\|^p \, \mathrm{d}\mu < \infty \right\}$$

and the semi-norm

$$\|f\|_{L^p} := \left(\int \|f\|^p \, \mathrm{d}\mu \right)^{\frac{1}{p}}, \quad f \in \mathcal{L}^p(\Omega, \mathcal{F}, \mu; X).$$

The space of all equivalence classes in $\mathcal{L}^p(\Omega, \mathcal{F}, \mu; X)$ with respect to $\| \ \|_{L^p}$ is denoted by $L^p(\Omega, \mathcal{F}, \mu; X) := L^p(\mu; X)$.

Claim: $L^1(\Omega, \mathcal{F}, \mu; X) = \overline{\mathcal{E}}$.

Step 2.a: $\left(L^1(\Omega, \mathcal{F}, \mu; X), \| \ \|_{L^1} \right)$ is complete.
    The proof is just a modification of the proof of the Fischer–Riesz theorem by the help of the following proposition.

Proposition A.1.3. *Let $(\Omega, \mathcal{F})$ be a measurable space and let X be a Banach space. Then:*

(i) *the set of $\mathcal{F}/\mathcal{B}(X)$-measurable functions from Ω to X is closed under the formation of pointwise limits, and*

(ii) *the set of strongly measurable functions from Ω to X is closed under the formation of pointwise limits.*

Proof. Simple exercise or see [Coh80, Proposition E.1, p. 350].   □

Step 2.b: $\mathcal{E}$ is a dense subset of $L^1(\Omega, \mathcal{F}, \mu; X)$ with respect to $\| \ \|_{L^1}$.
    This can be shown by the help of the following lemma.

Lemma A.1.4. *Let E be a metric space with metric d and let $f : \Omega \to E$ be strongly measurable. Then there exists a sequence f_n, $n \in \mathbb{N}$, of simple E-valued functions (i.e. f_n is $\mathcal{F}/\mathcal{B}(E)$-measurable and takes only a finite number of values) such that for arbitrary $\omega \in \Omega$ the sequence $d\big(f_n(\omega), f(\omega)\big)$, $n \in \mathbb{N}$, is monotonely decreasing to zero.*

Proof. [DPZ92, Lemma 1.1, p. 16] Let $\{e_k \mid k \in \mathbb{N}\}$ be a countable dense subset of $f(\Omega)$. For $m \in \mathbb{N}$ define

$$d_m(\omega) := \min\{d(f(\omega), e_k) \mid k \leqslant m\} \quad (= \operatorname{dist}(f(\omega), \{e_k, k \leqslant m\})),$$

$$k_m(\omega) := \min\{k \leqslant m \mid d_m(\omega) = d(f(\omega), e_k)\},$$

$$f_m(\omega) := e_{k_m(\omega)}.$$

Obviously f_m, $m \in \mathbb{N}$, are simple functions since they are $\mathcal{F}/\mathcal{B}(E)$-measurable (exercise) and

$$f_m(\Omega) \subset \{e_1, e_2, \ldots, e_m\}.$$

Moreover, by the density of $\{e_k \mid k \in \mathbb{N}\}$, the sequence $d_m(\omega)$, $m \in \mathbb{N}$, is monotonically decreasing to zero for arbitrary $\omega \in \Omega$. Since $d(f_m(\omega), f(\omega)) = d_m(\omega)$ the assertion follows. $\qquad\square$

Let now $f \in L^1(\mu; X)$. By the Lemma A.1.4 above we get the existence of a sequence of simple functions f_n, $n \in \mathbb{N}$, such that

$$\|f_n(\omega) - f(\omega)\| \downarrow 0 \quad \text{for all } \omega \in \Omega \text{ as } n \to \infty.$$

Hence $f_n \xrightarrow{n \to \infty} f$ in $\| \ \|_{L^1}$ by Lebesgue's dominated convergence theorem.

A.2. Properties of the Bochner integral

Proposition A.2.1 (Bochner inequality). *Let $f \in L^1(\Omega, \mathcal{F}, \mu; X)$. Then*

$$\left\| \int f \, d\mu \right\| \leqslant \int \|f\| \, d\mu.$$

Proof. We know the assertion is true for $f \in \mathcal{E}$, i.e. int $: \mathcal{E} \to X$ is linear, continuous with $\|\operatorname{int} f\| \leqslant \|f\|_{\mathcal{E}}$ for all $f \in \mathcal{E}$, so the same is true for its unique continuous extension $\overline{\operatorname{int}} : \overline{\mathcal{E}} = L^1(\mu; X) \to X$, i.e. for all $f \in L^1(X, \mu)$

$$\left\| \int f \, d\mu \right\| = \|\overline{\operatorname{int}} f\| \leqslant \|f\|_{\overline{\mathcal{E}}} = \int \|f\| \, d\mu. \qquad\square$$

Proposition A.2.2. *Let $f \in L^1(\Omega, \mathcal{F}, \mu; X)$. Then*

$$\int L \circ f \, d\mu = L\left(\int f \, d\mu \right)$$

holds for all $L \in L(X, Y)$, where Y is another Banach space.

Proof. Simple exercise or see [Coh80, Proposition E.11, p. 356]. $\qquad\square$

Proposition A.2.3 (Fundamental theorem of calculus). *Let* $-\infty < a < b < \infty$ *and* $f \in C^1([a,b];X)$. *Then*

$$f(t) - f(s) = \int_s^t f'(u) \, du := \begin{cases} \int 1_{[s,t]}(u) f'(u) \, du & \text{if } s \leqslant t \\ -\int 1_{[t,s]}(u) f'(u) \, du & \text{otherwise} \end{cases}$$

for all $s, t \in [a,b]$ *where* du *denotes the Lebesgue measure on* $\mathcal{B}(\mathbb{R})$.

Proof. Claim 1: If we set $F(t) = \int_s^t f'(u) \, du$, $t \in [a,b]$, we get that $F'(t) = f'(t)$ for all $t \in [a,b]$.

For that we have to prove that

$$\left\| \frac{1}{h} (F(t+h) - F(t)) - f'(t) \right\|_X \xrightarrow{h \to 0} 0.$$

To this end we fix $t \in [a,b]$ and take an arbitrary $\varepsilon > 0$. Since f' is continuous on $[a,b]$ there exists $\delta > 0$ such that $\|f'(u) - f'(t)\|_X < \varepsilon$ for all $u \in [a,b]$ with $|u - t| < \delta$. Then we obtain that

$$\left\| \frac{1}{h} (F(t+h) - F(t)) - f'(t) \right\|_X = \left\| \frac{1}{h} \int_t^{t+h} (f'(u) - f'(t)) \, du \right\|_X$$

$$\leqslant \frac{1}{h} \int_t^{t+h} \|f'(u) - f'(t)\|_X \, du < \varepsilon$$

if $t + h \in [a,b]$ and $|h| < \delta$.

Claim 2: If $\tilde{F} \in C^1([a,b];X)$ is a further function with $\tilde{F}' = F' = f'$ then there exists a constant $c \in X$ such that $F - \tilde{F} = c$.

For all $L \in X^* = L(X, \mathbb{R})$ we define $g_L := L(F - \tilde{F})$. Then $g'_L = 0$ and therefore g_L is constant. Since X^* separates the points of X by the Hahn–Banach theorem (see [Alt92, Satz 4.2, p. 114]) this implies that $F - \tilde{F}$ itself is constant. $\qquad\square$

B. Nuclear and Hilbert–Schmidt Operators

This chapter is identical to Chap. B in [FK01].

Let $(U, \langle \, , \, \rangle_U)$ and $(H, \langle \, , \, \rangle)$ be two separable Hilbert spaces. The space of all bounded linear operators from U to H is denoted by $L(U, H)$; for simplicity we write $L(U)$ instead of $L(U, U)$. If we speak of the adjoint operator of $L \in L(U, H)$ we write $L^* \in L(H, U)$. An element $L \in L(U)$ is called symmetric if $\langle Lu, v \rangle_U = \langle u, Lv \rangle_U$ for all $u, v \in U$. In addition, $L \in L(U)$ is called nonnegative if $\langle Lu, u \rangle \geqslant 0$ for all $u \in U$.

Definition B.0.1 (Nuclear operator). An element $T \in L(U, H)$ is said to be a nuclear operator if there exists a sequence $(a_j)_{j \in \mathbb{N}}$ in H and a sequence $(b_j)_{j \in \mathbb{N}}$ in U such that

$$Tx = \sum_{j=1}^{\infty} a_j \langle b_j, x \rangle_U \quad \text{for all } x \in U$$

and

$$\sum_{j \in \mathbb{N}} \|a_j\| \cdot \|b_j\|_U < \infty.$$

The space of all nuclear operators from U to H is denoted by $L_1(U, H)$.
If $U = H$, $T \in L_1(U, H)$ is nonnegative and symmetric, then T is called *trace class*.

Proposition B.0.2. *The space $L_1(U, H)$ endowed with the norm*

$$\|T\|_{L_1(U, H)} := \inf \left\{ \sum_{j \in \mathbb{N}} \|a_j\| \cdot \|b_j\|_U \;\middle|\; Tx = \sum_{j=1}^{\infty} a_j \langle b_j, x \rangle_U, \; x \in U \right\}$$

is a Banach space.

Proof. [MV92, Corollar 16.25, p. 154]. $\qquad\square$

Definition B.0.3. Let $T \in L(U)$ and let e_k, $k \in \mathbb{N}$, be an orthonormal basis of U. Then we define

$$\operatorname{tr} T := \sum_{k \in \mathbb{N}} \langle Te_k, e_k \rangle_U$$

if the series is convergent.

One has to notice that this definition could depend on the choice of the orthonormal basis. But there is the following result concerning nuclear operators.

Remark B.0.4. *If $T \in L_1(U)$ then $\operatorname{tr} T$ is well-defined independently of the choice of the orthonormal basis e_k, $k \in \mathbb{N}$. Moreover we have that*

$$|\operatorname{tr} T| \leqslant \|T\|_{L_1(U)}.$$

Proof. Let $(a_j)_{j \in \mathbb{N}}$ and $(b_j)_{j \in \mathbb{N}}$ be sequences in U such that

$$Tx = \sum_{j \in \mathbb{N}} a_j \langle b_j, x \rangle_U$$

for all $x \in U$ and $\sum_{j \in \mathbb{N}} \|a_j\|_U \cdot \|b_j\|_U < \infty$.

Then we get for any orthonormal basis e_k, $k \in \mathbb{N}$, of U that

$$\langle Te_k, e_k \rangle_U = \sum_{j \in \mathbb{N}} \langle e_k, a_j \rangle_U \cdot \langle e_k, b_j \rangle_U$$

and therefore

$$\sum_{k \in \mathbb{N}} |\langle Te_k, e_k \rangle_U| \leqslant \sum_{j \in \mathbb{N}} \sum_{k \in \mathbb{N}} |\langle e_k, a_j \rangle_U \cdot \langle e_k, b_j \rangle_U|$$

$$\leqslant \sum_{j \in \mathbb{N}} \left(\sum_{k \in \mathbb{N}} |\langle e_k, a_j \rangle_U|^2 \right)^{\frac{1}{2}} \cdot \left(\sum_{k \in \mathbb{N}} |\langle e_k, b_j \rangle_U|^2 \right)^{\frac{1}{2}}$$

$$= \sum_{j \in \mathbb{N}} \|a_j\|_U \cdot \|b_j\|_U < \infty.$$

This implies that we can exchange the summation to get that

$$\sum_{k \in \mathbb{N}} \langle Te_k, e_k \rangle_U = \sum_{j \in \mathbb{N}} \sum_{k \in \mathbb{N}} \langle e_k, a_j \rangle_U \cdot \langle e_k, b_j \rangle_U = \sum_{j \in \mathbb{N}} \langle a_j, b_j \rangle_U,$$

and the assertion follows.   □

Definition B.0.5 (Hilbert–Schmidt operator). A bounded linear operator $T : U \to H$ is called Hilbert–Schmidt if

$$\sum_{k \in \mathbb{N}} \|Te_k\|^2 < \infty$$

where e_k, $k \in \mathbb{N}$, is an orthonormal basis of U.

The space of all Hilbert–Schmidt operators from U to H is denoted by $L_2(U, H)$.

Remark B.0.6.   *(i) The definition of Hilbert–Schmidt operator and the number*

$$\|T\|^2_{L_2(U,H)} := \sum_{k\in\mathbb{N}} \|Te_k\|^2$$

does not depend on the choice of the orthonormal basis e_k, $k \in \mathbb{N}$, and we have that $\|T\|_{L_2(U,H)} = \|T^*\|_{L_2(H,U)}$. For simplicity we also write $\|T\|_{L_2}$ instead of $\|T\|_{L_2(U,H)}$.

(ii) $\|T\|_{L(U,H)} \leqslant \|T\|_{L_2(U,H)}$.

(iii) Let G be another Hilbert space and $S_1 \in L(H,G)$, $S_2 \in L(G,U)$, $T \in L_2(U,H)$. Then $S_1T \in L_2(U,G)$ and $TS_2 \in L_2(G,H)$ and

$$\|S_1T\|_{L_2(U,G)} \leqslant \|S_1\|_{L(H,G)}\|T\|_{L_2(U,H)},$$

$$\|TS_2\|_{L_2(G,H)} \leqslant \|T\|_{L(U,H)}\|S_2\|_{L_2(G,U)}.$$

Proof.   (i) If e_k, $k \in \mathbb{N}$, is an orthonormal basis of U and f_k, $k \in \mathbb{N}$, is an orthonormal basis of H we obtain by the Parseval identity that

$$\sum_{k\in\mathbb{N}}\|Te_k\|^2 = \sum_{k\in\mathbb{N}}\sum_{j\in\mathbb{N}}|\langle Te_k, f_j\rangle|^2 = \sum_{j\in\mathbb{N}}\|T^*f_j\|^2_U$$

and therefore the assertion follows.

(ii) Let $x \in U$ and f_k, $k \in \mathbb{N}$, be an orthonormal basis of H. Then we get that

$$\|Tx\|^2 = \sum_{k\in\mathbb{N}}\langle Tx, f_k\rangle^2 \leqslant \|x\|^2_U \sum_{k\in\mathbb{N}}\|T^*f_k\|^2_U = \|T\|^2_{L_2(U,H)} \cdot \|x\|^2_U.$$

(iii) Let e_k, $k \in \mathbb{N}$ be an orthonormal basis of U. Then

$$\sum_{k\in\mathbb{N}}\|S_1Te_k\|^2_G \leqslant \|S_1\|^2_{L(H,G)}\|T\|^2_{L_2(U,H)}.$$

Furthermore, since $(TS_2)^* = S_2^*T^*$, it follows that by the above and (i) that $TS_2 \in L_2(G,H)$ and

$$\begin{aligned}
\|TS_2\|_{L_2(G,H)} &= \|(TS_2)^*\|_{L_2(H,G)} \\
&= \|S_2^*T^*\|_{L_2(H,G)} \\
&\leqslant \|S_2\|_{L(G,U)} \cdot \|T\|_{L_2(U,H)}.
\end{aligned}$$

$\square$

Proposition B.0.7. *Let* $S, T \in L_2(U, H)$ *and let* e_k, $k \in \mathbb{N}$, *be an orthonormal basis of* U. *If we define*

$$\langle T, S \rangle_{L_2} := \sum_{k \in \mathbb{N}} \langle Se_k, Te_k \rangle$$

we obtain that $(L_2(U, H), \langle\ , \ \rangle_{L_2})$ *is a separable Hilbert space.*

If f_k, $k \in \mathbb{N}$, *is an orthonormal basis of* H *we get that* $f_j \otimes e_k := f_j \langle e_k, \cdot \rangle_U$, $j, k \in \mathbb{N}$, *is an orthonormal basis of* $L_2(U, H)$.

Proof. We have to prove the completeness and the separability.

1. $L_2(U, H)$ is complete:

   Let T_n, $n \in \mathbb{N}$, be a Cauchy sequence in $L_2(U, H)$. Then it is clear that it is also a Cauchy sequence in $L(U, H)$. Because of the completeness of $L(U, H)$ there exists an element $T \in L(U, H)$ such that $\|T_n - T\|_{L(U,H)} \longrightarrow 0$ as $n \to \infty$. But by the lemma of Fatou we also have for any orthonormal basis e_k, $k \in \mathbb{N}$, of U that

$$\|T_n - T\|_{L_2}^2 = \sum_{k \in \mathbb{N}} \langle (T_n - T)e_k, (T_n - T)e_k \rangle$$

$$= \sum_{k \in \mathbb{N}} \liminf_{m \to \infty} \big\|(T_n - T_m)e_k\big\|^2$$

$$\leqslant \liminf_{m \to \infty} \sum_{k \in \mathbb{N}} \big\|(T_n - T_m)e_k\big\|^2 = \liminf_{m \to \infty} \|T_n - T_m\|_{L_2}^2 < \varepsilon$$

   for all $n \in \mathbb{N}$ big enough. Therefore the assertion follows.

2. $L_2(U, H)$ is separable:

   If we define $f_j \otimes e_k := f_j \langle e_k, \cdot \rangle_U$, $j, k \in \mathbb{N}$, then it is clear that $f_j \otimes e_k \in L_2(U, H)$ for all $j, k \in \mathbb{N}$ and for arbitrary $T \in L_2(U, H)$ we get that

$$\langle f_j \otimes e_k, T \rangle_{L_2} = \sum_{n \in \mathbb{N}} \langle e_k, e_n \rangle_U \cdot \langle f_j, Te_n \rangle = \langle f_j, Te_k \rangle.$$

   Therefore it is obvious that $f_j \otimes e_k$, $j, k \in \mathbb{N}$, is an orthonormal system. In addition, $T = 0$ if $\langle f_j \otimes e_k, T \rangle_{L_2} = 0$ for all $j, k \in \mathbb{N}$, and therefore $\mathrm{span}(f_j \otimes e_k \mid j, k \in \mathbb{N})$ is a dense subspace of $L_2(U, H)$. $\square$

Proposition B.0.8. *Let* $(G, \langle\ , \ \rangle_G)$ *be a further separable Hilbert space. If* $T \in L_2(U, H)$ *and* $S \in L_2(H, G)$ *then* $ST \in L_1(U, G)$ *and*

$$\|ST\|_{L_1(U,G)} \leqslant \|S\|_{L_2} \cdot \|T\|_{L_2}.$$

Proof. Let f_k, $k \in \mathbb{N}$, be an orthonormal basis of H. Then we have that

$$STx = \sum_{k \in \mathbb{N}} \langle Tx, f_k \rangle Sf_k, \quad x \in U$$

and therefore

$$\|ST\|_{L_1(U,G)} \leqslant \sum_{k\in\mathbb{N}} \|T^* f_k\|_U \cdot \|Sf_k\|_G$$

$$\leqslant \left(\sum_{k\in\mathbb{N}} \|T^* f_k\|_U^2\right)^{\frac{1}{2}} \cdot \left(\sum_{k\in\mathbb{N}} \|Sf_k\|_G^2\right)^{\frac{1}{2}} = \|S\|_{L_2} \cdot \|T\|_{L_2}. \qquad \square$$

Remark B.0.9. *Let e_k, $k \in \mathbb{N}$, be an orthonormal basis of U. If $T \in L(U)$ is symmetric, nonnegative with $\sum_{k\in\mathbb{N}} \langle Te_k, e_k \rangle_U < \infty$ then $T \in L_1(U)$.*

Proof. The result is obvious by the previous proposition and the fact that there exists $T^{\frac{1}{2}} \in L(U)$ nonnegative and symmetric such that $T = T^{\frac{1}{2}}T^{\frac{1}{2}}$ (see Proposition 2.3.4). Then $T^{\frac{1}{2}} \in L_2(U)$. $\qquad \square$

Proposition B.0.10. *Let $L \in L(H)$ and $B \in L_2(U,H)$. Then $LBB^* \in L_1(H)$, $B^*LB \in L_1(U)$ and we have that*

$$\operatorname{tr} LBB^* = \operatorname{tr} B^*LB.$$

Proof. We know by Remark B.0.6 (iii) and Proposition B.0.8 that $LBB^* \in L_1(H)$ and $B^*LB \in L_1(U)$. Let e_k, $k \in \mathbb{N}$, be an orthonormal basis of U and let f_k, $k \in \mathbb{N}$, be an orthonormal basis of H. Then the Parseval identity implies that

$$\sum_{k\in\mathbb{N}}\sum_{n\in\mathbb{N}} |\langle f_k, Be_n \rangle \cdot \langle f_k, LBe_n \rangle|$$

$$\leqslant \sum_{n\in\mathbb{N}} \left(\sum_{k\in\mathbb{N}} |\langle f_k, Be_n \rangle|^2\right)^{\frac{1}{2}} \cdot \left(\sum_{k\in\mathbb{N}} |\langle f_k, LBe_n \rangle|^2\right)^{\frac{1}{2}}$$

$$= \sum_{n\in\mathbb{N}} \|Be_n\| \cdot \|LBe_n\| \leqslant \|L\|_{L(H)} \cdot \|B\|_{L_2}^2.$$

Therefore, it is allowed to interchange the sums to obtain that

$$\operatorname{tr} LBB^* = \sum_{k\in\mathbb{N}} \langle LBB^* f_k, f_k \rangle = \sum_{k\in\mathbb{N}} \langle B^* f_k, B^* L^* f_k \rangle_U$$

$$= \sum_{k\in\mathbb{N}}\sum_{n\in\mathbb{N}} \langle B^* f_k, e_n \rangle_U \cdot \langle B^* L^* f_k, e_n \rangle_U = \sum_{n\in\mathbb{N}}\sum_{k\in\mathbb{N}} \langle f_k, Be_n \rangle \cdot \langle f_k, LBe_n \rangle$$

$$= \sum_{n\in\mathbb{N}} \langle Be_n, LBe_n \rangle = \sum_{n\in\mathbb{N}} \langle e_n, B^* LBe_n \rangle_U = \operatorname{tr} B^*LB. \qquad \square$$

C. Pseudo Inverse of Linear Operators

This chapter is a slight modification of Chapter C in [FK01].

Let $(U, \langle\,,\,\rangle_U)$ and $(H, \langle\,,\,\rangle)$ be two Hilbert spaces.

Definition C.0.1 (Pseudo inverse). Let $T \in L(U, H)$ and $\text{Ker}(T) := \{x \in U \mid Tx = 0\}$. The pseudo inverse of T is defined as

$$T^{-1} := \left(T|_{\text{Ker}(T)^\perp}\right)^{-1} : T\left(\text{Ker}(T)^\perp\right) = T(U) \to \text{Ker}(T)^\perp.$$

(Note that T is one-to-one on $\text{Ker}(T)^\perp$.)

Remark C.0.2. *(i) There is an equivalent way of defining the pseudo inverse of a linear operator $T \in L(U, H)$. For $x \in T(U)$ one sets $T^{-1}x \in U$ to be the solution of minimal norm of the equation $Ty = x$, $y \in U$.*

(ii) If $T \in L(U, H)$ then $T^{-1} : T(U) \to \text{Ker}(T)^\perp$ is linear and bijective.

Proposition C.0.3. *Let $T \in L(U)$ and T^{-1} the pseudo inverse of T.*

(i) If we define an inner product on $T(U)$ by

$$\langle x, y \rangle_{T(U)} := \langle T^{-1}x, T^{-1}y \rangle_U \quad \text{for all } x, y \in T(U),$$

then $\left(T(U), \langle\,,\,\rangle_{T(U)}\right)$ is a Hilbert space.

(ii) Let e_k, $k \in \mathbb{N}$, be an orthonormal basis of $(\text{Ker}\,T)^\perp$. Then Te_k, $k \in \mathbb{N}$, is an orthonormal basis of $\left(T(U), \langle\,,\,\rangle_{T(U)}\right)$.

Proof. $T : (\text{Ker}\,T)^\perp \to T(U)$ is bijective and an isometry if $(\text{Ker}\,T)^\perp$ is equipped with $\langle\,,\,\rangle_U$ and $T(U)$ with $\langle\,,\,\rangle_{T(U)}$. $\qquad\square$

Now we want to present a result about the images of linear operators. To this end we need the following lemma.

Lemma C.0.4. *Let $T \in L(U, H)$. Then the set $\overline{TB_c(0)}$ $(= \{Tu \mid u \in U, \|u\|_U \leqslant c\})$, $c \geqslant 0$, is convex and closed.*

Proof. Since T is linear it is obvious that the set is convex.

Since a convex subset of a Hilbert space is closed (with respect to the norm) if and only if it is weakly closed, it suffices to show that $\overline{TB_c(0)}$ is weakly closed. Since $T : U \to H$ is linear and continuous (with respect to the norms on U, H respectively) it is also obviously continuous with respect to the weak topologies on U, H respectively. But by the Banach–Alaoglou theorem (see e.g. [RS72, Theorem IV.21, p. 115]) closed balls in a Hilbert space are weakly compact. Hence $\overline{B_c(0)}$ is weakly compact, and so is its continuous image, i.e. $T\overline{B_c(0)}$ is weakly compact, therefore weakly closed. $\qquad\square$

Proposition C.0.5. *Let $(U_1, \langle\,,\,\rangle_1)$ and $(U_2, \langle\,,\,\rangle_2)$ be two Hilbert spaces. In addition, we take $T_1 \in L(U_1, H)$ and $T_2 \in L(U_2, H)$. Then the following statements hold.*

(i) If there exists a constant $c \geqslant 0$ such that $\|T_1^* x\|_1 \leqslant c\|T_2^* x\|_2$ for all $x \in H$ then $\{T_1 u \mid u \in U_1, \|u\|_1 \leqslant 1\} \subset \{T_2 v \mid v \in U_2, \|v\|_2 \leqslant c\}$. In particular, this implies that $\operatorname{Im} T_1 \subset \operatorname{Im} T_2$.

(ii) If $\|T_1^* x\|_1 = \|T_2^* x\|_2$ for all $x \in H$ then $\operatorname{Im} T_1 = \operatorname{Im} T_2$ and $\|T_1^{-1} x\|_1 = \|T_2^{-1} x\|_2$ for all $x \in \operatorname{Im} T_1$.

Proof. [DPZ92, Proposition B.1, p. 407]

(i) Assume that there exists $u_0 \in U_1$ such that

$$\|u_0\|_1 \leqslant 1 \quad \text{and} \quad T_1 u_0 \notin \{T_2 v \mid v \in U_2, \|v\|_2 \leqslant c\}.$$

By Lemma C.0.4 we know that the set $\{T_2 v \mid v \in U_2, \|v\|_2 \leqslant c\}$ is closed and convex. Therefore, we get by the separation theorem (see [Alt92, 5.11 Trennungssatz, p. 166]) there exists $x \in H$, $x \neq 0$, such that

$$1 < \langle x, T_1 u_0 \rangle \quad \text{and} \quad \langle x, T_2 v \rangle \leqslant 1 \text{ for all } v \in U_2 \text{ with } \|v\|_2 \leqslant c.$$

Thus $\|T_1^* x\|_1 > 1$ and $c\|T_2^* x\|_2 = \sup_{\|v\|_2 \leqslant c} |\langle T_2^* x, v \rangle_2| \leqslant 1$, a contradiction.

(ii) By (i) we know that $\operatorname{Im} T_1 = \operatorname{Im} T_2$. It remains to verify that

$$\|T_1^{-1} x\|_1 = \|T_2^{-1} x\|_2 \quad \text{for all } x \in \operatorname{Im} T_1.$$

If $x = 0$ then $\|T_1^{-1} 0\|_1 = 0 = \|T_2^{-1} 0\|_2$.

If $x \in \operatorname{Im} T_1 \setminus \{0\}$ then there exist $u_1 \in (\operatorname{Ker} T_1)^\perp$ and $u_2 \in (\operatorname{Ker} T_2)^\perp$ such that $x = T_1 u_1 = T_2 u_2$. We have to show that $\|u_1\|_1 = \|u_2\|_2$.

Assume that $\|u_1\|_1 > \|u_2\|_2 > 0$. Then (i) implies that

$$\frac{x}{\|u_2\|_2} = T_2\left(\frac{u_2}{\|u_2\|_2}\right)$$

$$\in \{T_2 v \mid v \in U_2, \|v\|_2 \leqslant 1\} = \{T_1 u \mid u \in U_1, \|u\|_1 \leqslant 1\}.$$

But

$$\frac{x}{\|u_2\|_2} = T_1\left(\frac{u_1}{\|u_2\|_2}\right) \quad \text{and} \quad \left\|\frac{u_1}{\|u_2\|_2}\right\|_1 > 1,$$

therefore, there exists $\tilde{u}_1 \in U_1$, $\|\tilde{u}_1\|_1 \leqslant 1$, so that for $\tilde{u}_2 := \frac{u_1}{\|u_2\|_2} \in$ $(\operatorname{Ker} T_1)^{\perp}$ we have

$$T_1 \tilde{u}_1 = \frac{x}{\|u_2\|_2} = T_1 \tilde{u}_2 \,, \quad \text{i.e. } \tilde{u}_1 - \tilde{u}_2 \in \operatorname{Ker} T_1.$$

Therefore,

$$0 = \langle \tilde{u}_1 - \tilde{u}_2, \tilde{u}_2 \rangle_1 = \langle \tilde{u}_1, \tilde{u}_2 \rangle_1 - \|\tilde{u}_2\|_1^2$$
$$\leqslant \|\tilde{u}_1\|_1 \|\tilde{u}_2\|_1 - \|\tilde{u}_2\|_1^2 = \left(1 - \|\tilde{u}_2\|_1\right) \|\tilde{u}_2\|_1.$$

This is a contradiction.   □

Corollary C.0.6. *Let $T \in L(U, H)$ and set $Q := TT^* \in L(H)$. Then we have*

$$\operatorname{Im} Q^{\frac{1}{2}} = \operatorname{Im} T \quad \text{and} \quad \left\| Q^{-\frac{1}{2}} x \right\| = \|T^{-1}x\|_U \ \text{for all } x \in \operatorname{Im} T,$$

where $Q^{-\frac{1}{2}}$ is the pseudo inverse of $Q^{\frac{1}{2}}$.

Proof. Since by Lemma 2.3.4 $Q^{\frac{1}{2}}$ is symmetric we have for all $x \in H$ that

$$\left\| \left(Q^{\frac{1}{2}}\right)^* x \right\|^2 = \left\| Q^{\frac{1}{2}} x \right\|^2 = \langle Qx, x \rangle = \langle TT^*x, x \rangle = \|T^*x\|_U^2.$$

Therefore the assertion follows by Proposition C.0.5.   □

D. Some Tools from Real Martingale Theory

We need the following Burkholder–Davis inequality for real-valued continuous local martingales.

Proposition D.0.1. *Let $(N_t)_{t\in[0,T]}$ be a real-valued continuous local martingale on a probability space (Ω, E, P) with respect to a normal filtration $(\mathcal{F}_t)_{t\in[0,T]}$. Then for all stopping times $\tau(\leqslant T)$*

$$E(\sup_{t\in[0,\tau]} |N_t|) \leqslant 3E(\langle N\rangle_\tau^{1/2}).$$

Proof. See e.g. [KS88, Theorem 3.28]. $\qquad\square$

Corollary D.0.2. *Let $\varepsilon, \delta \in]0, \infty[$. Then for N as in Proposition D.0.1*

$$P(\sup_{t\in[0,T]} |N_t| \geqslant \varepsilon) \leqslant \frac{3}{\varepsilon}E(\langle N\rangle_T^{1/2} \wedge \delta) + P(\langle N\rangle_T^{1/2} > \delta).$$

Proof. Let

$$\tau := \inf\{t \geqslant 0|\ \langle N\rangle_t^{1/2} > \delta\} \wedge T.$$

Then $\tau(\leqslant T)$ is an $\mathcal{F}_t$-stopping time. Hence by Proposition D.0.1

$$P\left(\sup_{t\in[0,T]} |N_t| \geqslant \varepsilon\right)$$

$$=P\left(\sup_{t\in[0,T]} |N_t| \geqslant \varepsilon, \tau = T\right) + P\left(\sup_{t\in[0,T]} |N_t| \geqslant \varepsilon, \tau < T\right)$$

$$\leqslant\frac{3}{\varepsilon}E(\langle N\rangle_\tau^{1/2}) + P\left(\sup_{t\in[0,T]} |N_t| \geqslant \varepsilon, \langle N\rangle_T^{1/2} > \delta\right)$$

$$\leqslant\frac{3}{\varepsilon}E(\langle N\rangle_T^{1/2} \wedge \delta) + P(\langle N\rangle_T^{1/2} > \delta).$$

$\qquad\square$

E. Weak and Strong Solutions: the Yamada-Watanabe Theorem

Let $(\Omega, \mathcal{F}, P)$ be a complete probability space with normal filtration $\mathcal{F}_t$, $t \in [0, \infty[$. Below we shall call $((\Omega, \mathcal{F}, P, (\mathcal{F}_t))$ a *stochastic basis*. Let $d, d_1 \in \mathbb{N}$ and let $M(d \times d_1, \mathbb{R})$ denote the set of all real $d \times d_1$-matrices equipped with the norm (3.1.2). Let

$$W^d := C([0, \infty[\to \mathbb{R}^d) \tag{E.0.1}$$

and

$$W_0^d := \{w \in W^d | w(0) = 0\}. \tag{E.0.2}$$

W^d is equipped with metric

$$\varrho(w_1, w_2) := \sum_{k=1}^{\infty} 2^{-k} \big(\max_{0 \leqslant t \leqslant k} |w_1(t) - w_2(t)| \wedge 1 \big), \quad w_1, w_2 \in W^d, \tag{E.0.3}$$

which makes it a Polish space. Its Borel σ-algebra is denoted by $\mathcal{B}(W^d)$. Let $\mathcal{B}_t(W^d)$ denote the σ-Algebra generated by all maps π_s, $0 \leqslant s \leqslant t$, where $\pi_s(w) := w(s)$, $w \in W^d$. Let $\mathcal{A}^{d,d_1}$ denote the set of all $\mathcal{B}([0, \infty[) \otimes \mathcal{B}(W^d)/\mathcal{B}(M(d \times d_1, \mathbb{R}))$-measurable maps $\alpha : [0, \infty[\times W^d \to M(d \times d_1, \mathbb{R})$ such that for each $t \in [0, \infty[$ the map

$$W^d \ni w \mapsto \alpha(t, w) \in M(d \times d_1, \mathbb{R})$$

is $\mathcal{B}_t(W^d)/\mathcal{B}(M(d \times d_1, \mathbb{R}))$-measurable.

E.1. The main result

Fix $\sigma \in \mathcal{A}^{d,d_1}$ and $b \in \mathcal{A}^{d,1}$ and consider the following stochastic differential equation:

$$dX(t) = b(t, X) \, dt + \sigma(t, X) \, dW(t), \quad t \in [0, \infty[. \tag{E.1.1}$$

Definition E.1.1. An $\mathbb{R}^d$-valued continuous, $(\mathcal{F}_t)$-adapted process $X(t)$, $t \in [0, \infty[$, on some stochastic basis $(\Omega, \mathcal{F}, P, (\mathcal{F}_t))$ is called a *(weak) solution* to (E.1.1), if

(i)

$$\int_0^t |b(s, X)| \, ds < \infty \quad P\text{-a.e. for all } t \in [0, \infty[.$$

(ii)

$$\int_0^t \|\sigma(s, X)\|^2 \, ds < \infty \quad P\text{-a.e. for all } t \in [0, \infty[.$$

(iii) There exists an $\mathbb{R}^{d_1}$-valued standard $(\mathcal{F}_t)$-Wiener process $W(t)$, $t \in [0, \infty[$, on $(\Omega, \mathcal{F}, P)$ such that P-a.e.

$$X(t) = X(0) + \int_0^t b(s, X) \, ds + \int_0^t \sigma(s, X) \, dW(s), \quad t \in [0, \infty[. \quad \text{(E.1.2)}$$

Remark E.1.2. *(i) Clearly, by the measurability assumption on elements in $\mathcal{A}^{d,d_1}$ it follows that if X is a solution, then $[0, t] \times \Omega \ni (s, \omega) \mapsto \sigma(s, X(\omega))$ is $\mathcal{B}([0, t]) \otimes \mathcal{F}/\mathcal{B}(M(d \times d_1, \mathbb{R}))$-measurable and $\sigma(t, X)$ is $\mathcal{F}_t$-measurable for $t \in [0, \infty[$. Likewise for $b(\cdot, X)$. The $(\mathcal{F}_t)$-adaptedness for $\sigma(\cdot, X)$ and $b(\cdot, X)$ follows since the $(\mathcal{F}_t)$-adaptiveness of X is equivalent to the $\mathcal{F}_t/\mathcal{B}_t(W^d)$ measurability of X.*

(ii) Below we shall briefly say (X, W) in Definition E.1.1 is a (weak) solution to (E.1.1) not always mentioning explicitly the stochastic basis, that comes with it.

Definition E.1.3. We say that *(weak) uniqueness* holds for (E.1.1) if whenever X and X' are two (weak) solutions (with stochastic bases $(\Omega, \mathcal{F}, P, (\mathcal{F}_t))$, $(\Omega', \mathcal{F}', P', (\mathcal{F}'_t))$ and associated Wiener processes $W(t)$, $W'(t)$, $t \in [0, \infty[)$ such that

$$P \circ X(0)^{-1} = P' \circ X'(0)^{-1},$$

(as measures on $(\mathbb{R}^d, \mathcal{B}(\mathbb{R}^d)))$), then

$$P \circ X^{-1} = P' \circ (X')^{-1}$$

(as measures on $(W^d, \mathcal{B}(W^d)))$).

Definition E.1.4. We say that *pathwise uniqueness* holds for (E.1.1), if whenever X and X' are two (weak) solutions on the same stochastic basis $(\Omega, \mathcal{F}, P, (\mathcal{F}_t))$ and with the same $(\mathcal{F}_t)$-Wiener process $W(t)$, $t \in [0, \infty[$ on $(\Omega, \mathcal{F}, P)$ such that $X(0) = X'(0)$ P-a.e., then P-a.e.

$$X(t) = X'(t), t \in [0, \infty[.$$

To define strong solutions we need to introduce the following class $\hat{\mathcal{E}}$ of maps:

Let $\hat{\mathcal{E}}$ denote the set of all maps $F : \mathbb{R}^d \times W_0^{d_1} \to W^d$ such that for every probability measure μ on $(\mathbb{R}^d, \mathcal{B}(\mathbb{R}^d))$ there exists a $\overline{\mathcal{B}(\mathbb{R}^d) \otimes \mathcal{B}(W_0^{d_1})}^{\mu \otimes P^W} / \mathcal{B}(W^d)$-measurable map $F_\mu : \mathbb{R}^d \times W_0^{d_1} \to W^d$ such that for μ-a.e. $x \in \mathbb{R}^d$

$$F(x, w) = F_\mu(x, w) \text{ for } P^W\text{-a.e. } w \in W_0^{d_1}.$$

Here $\overline{\mathcal{B}(\mathbb{R}^d) \otimes \mathcal{B}(W_0^{d_1})}^{\mu \otimes P^W}$ denotes the completion of $\mathcal{B}(\mathbb{R}^d) \otimes \mathcal{B}(W_0^{d_1})$ with respect to $\mu \otimes P^W$, and P^W denotes classical Wiener measure on $(W_0^{d_1}, \mathcal{B}(W_0^{d_1}))$.

Let $F \in \hat{\mathcal{E}}$. For an $\mathcal{F}/\mathcal{B}(\mathbb{R}^d)$-measurable map $\xi : \Omega \to \mathbb{R}^d$ on some probability space $(\Omega, \mathcal{F}, P)$ and an $\mathbb{R}^{d_1}$-valued, standard Wiener process $W(t)$, $t \in [0, \infty[$, on $(\Omega, \mathcal{F}, P)$ independent of ξ, we set

$$F(\xi, W) := F_{P \circ \xi^{-1}}(\xi, W).$$

Definition E.1.5. A (weak) solution X to (E.1.1) on $(\Omega, \mathcal{F}, P, (\mathcal{F}_t))$ and associated Wiener process $W(t)$, $t \in [0, \infty[$, is called a *strong solution* if there exists $F \in \hat{\mathcal{E}}$ such that for $x \in \mathbb{R}^d$, $w \mapsto F(x, w)$ is $\overline{\mathcal{B}_t(W_0^{d_1})}^{P^W} / \mathcal{B}_t(W^d)$-measurable for every $t \in [0, \infty[$ and

$$X = F(X(0), W) \quad P\text{-a.e.},$$

where $\overline{\mathcal{B}_t(W_0^{d_1})}^{P^W}$ denotes the completion with respect to P^W in $\mathcal{B}(W_0^{d_1})$.

Definition E.1.6. Equation (E.1.1) is said to have a *unique strong solution*, if there exists $F \in \hat{\mathcal{E}}$ satisfying the adaptiveness condition in Definition E.1.5 and such that:

1. For every $\mathbb{R}^{d_1}$-valued standard $(\mathcal{F}_t)$-Wiener process $W(t)$, $t \in [0, \infty[$, on a stochastic basis $(\Omega, \mathcal{F}, P, (\mathcal{F}_t))$ and any $\mathcal{F}_0/\mathcal{B}(\mathbb{R}^d)$-measurable $\xi : \Omega \to \mathbb{R}^d$ the continuous process

$$X := F(\xi, W)$$

   satisfies (i), (ii) and (E.1.2) in Definition E.1.1, i.e. $(F(\xi, W), W)$ is a (weak) solution to (E.1.1), and $X(0) = \xi$ P-a.e..

2. For any (weak) solution (X, W) to (E.1.1) we have

$$X = F(X(0), W) \ P\text{-a.e.}.$$

Remark E.1.7. *Since $X(0)$ in the above definition is P-independent of W, thus*

$$P \circ (X(0), W)^{-1} = \mu \otimes P^W,$$

we have that the existence of a unique strong solution for (E.1.1) implies that also (weak) uniqueness holds.

Now we can formulate the main result of this section.

Theorem E.1.8. *Let $\sigma \in \mathcal{A}^{d,d_1}$ and $b \in \mathcal{A}^{d,1}$. Then equation (E.1.1) has a unique strong solution if and only if both of the following properties hold:*

(i) For every probability measure μ on $(\mathbb{R}^d, \mathcal{B}(\mathbb{R}^d))$ there exists a (weak) solution (X, W) of (E.1.1) such that μ is the distribution of $X(0)$.

(ii) Pathwise uniqueness holds for (E.1.1).

Proof. Suppose (E.1.1) has a unique strong solution. Then (ii) obviously holds. To show (i) one only has to take the classical Wiener space $(W_0^{d_1}, \mathcal{B}(W_0^{d_1}), P^W)$ and consider $(\mathbb{R}^d \times W_0^{d_1}, \overline{\mathcal{B}(\mathbb{R}^d) \otimes \mathcal{B}(W_0^{d_1})}^{\mu \otimes P^W}, \mu \otimes P^W)$ with filtration

$$\bigcap_{\varepsilon > 0} \sigma(\mathcal{B}(\mathbb{R}^d) \otimes \mathcal{B}_{t+\varepsilon}(W_0^{d_1}), \mathcal{N}), \quad t \geq 0,$$

where $\mathcal{N}$ denotes all $\mu \otimes P^W$-zero sets in $\overline{\mathcal{B}(\mathbb{R}^d) \otimes \mathcal{B}(W_0^{d_1})}^{\mu \otimes P^W}$. Let $\xi : \mathbb{R}^d \times W_0^{d_1} \to \mathbb{R}^d$ and $W : \mathbb{R}^d \times W_0^{d_1} \to W_0^{d_1}$ be the canonical projections. Then $X := F(\xi, W)$ is the desired weak solution in (i).

Now let us suppose that (i) and (ii) hold. The proof that then there exists a unique strong solution for (E.1.1) is quite technical. We structure it through a series of lemmas.

Lemma E.1.9. *Let $(\Omega, \mathcal{F})$ be a measurable space such that $\{\omega\} \in \mathcal{F}$ for all $\omega \in \Omega$ and such that*

$$D := \{(\omega, \omega) | \omega \in \Omega\} \in \mathcal{F} \otimes \mathcal{F}$$

(which is e.g. the case if Ω is a Polish space and $\mathcal{F}$ its Borel σ-algebra). Let P_1, P_2 be probability measures on $(\Omega, \mathcal{F})$ such that $P_1 \otimes P_2(D) = 1$. Then $P_1 = P_2 = \delta_{\omega_0}$ for some $\omega_0 \in \Omega$.

Proof. Let $f : \Omega \to [0, \infty[$ be $\mathcal{F}$-measurable. Then

$$\int f(\omega_1) P_1(d\omega_1) = \iint f(\omega_1) P_1(d\omega_1) P_2(d\omega_2)$$

$$= \iint 1_D(\omega_1, \omega_2) f(\omega_1) P_1(d\omega_1) P_2(d\omega_2)$$

$$= \iint 1_D(\omega_1, \omega_2) f(\omega_2) P_1(d\omega_1) P_2(d\omega_2) = \int f(\omega_2) P_2(d\omega_2),$$

so $P_1 = P_2$. Furthermore,

$$1 = \iint 1_D(\omega_1, \omega_2) P_1(d\omega_1) P_2(d\omega_2) = \int P_1(\{\omega_2\}) P_2(d\omega_2),$$

hence $1 = P_1(\{\omega_2\})$ for P_2-a.e. $\omega_2 \in \Omega$. Therefore, $P_1 = \delta_{\omega_0}$ for some $\omega_0 \in \Omega$. $\qquad\square$

Fix a probability measure μ on $(\mathbb{R}^d, \mathcal{B}(\mathbb{R}^d))$ and let (X, W) with stochastic basis $(\Omega, \mathcal{F}, P, (\mathcal{F}_t))$ be a (weak) solution to (E.1.1) with initial distribution μ. Define a probability measure P_μ on $(\mathbb{R}^d \times W^d \times W_0^{d_1}, \mathcal{B}(\mathbb{R}^d) \otimes \mathcal{B}(W^d) \otimes \mathcal{B}(W_0^{d_1}))$, by

$$P_\mu := P \circ (X(0), X, W)^{-1}.$$

Lemma E.1.10. *There exists a family $K_\mu((x, w), dw_1), x \in \mathbb{R}^d, w \in W_0^{d_1}$, of probability measures on $(W^d, \mathcal{B}(W^d))$ having the following properties:*

(i) For every $A \in \mathcal{B}(W^d)$ the map

$$\mathbb{R}^d \times W_0^{d_1} \ni (x, w) \mapsto K_\mu((x, w), A)$$

is $\mathcal{B}(\mathbb{R}^d) \otimes \mathcal{B}(W_0^{d_1})$-measurable.

(ii) For every $\mathcal{B}(\mathbb{R}^d) \otimes \mathcal{B}(W^d) \otimes \mathcal{B}(W_0^{d_1})$-measurable map $f : \mathbb{R}^d \times W^d \times W_0^{d_1} \to [0, \infty[$ we have

$$\int f(x, w_1, w) P_\mu(dx\, dw_1\, dw)$$

$$= \int_{\mathbb{R}^d} \int_{W_0^{d_1}} \int_{W^d} f(x, w_1, w) K_\mu((x, w), dw_1) P^W(dw) \mu(dx).$$

(iii) If $t \in [0, \infty[$ and $f : W^d \to [0, \infty[$ is $\mathcal{B}_t(W^d)$-measurable, then

$$\mathbb{R}^d \times W_0^{d_1} \ni (x, w) \mapsto \int f(w_1) K_\mu((x, w), dw_1)$$

is $\overline{\mathcal{B}(\mathbb{R}^d) \otimes \mathcal{B}_t(W_0^{d_1})}^{\mu \otimes P^W}$-measurable, where $\overline{\mathcal{B}(\mathbb{R}^d) \otimes \mathcal{B}_t(W_0^{d_1})}^{\mu \otimes P^W}$ denotes the completion with respect to $\mu \otimes P^W$ in $\mathcal{B}(\mathbb{R}^d) \otimes \mathcal{B}(W_0^{d_1})$.

Proof. Let $\Pi : \mathbb{R}^d \times W^d \times W_0^{d_1} \to \mathbb{R}^d \times W_0^{d_1}$ be the canonical projection. Since $X(0)$ is $\mathcal{F}_0$-measurable, hence P-independent of W, it follows that

$$P_\mu \circ \Pi^{-1} = P \circ (X(0), W)^{-1} = \mu \otimes P^W.$$

Hence by the existence result on regular conditional distributions (cf. e.g. [IW81, Corollary to Theorem 3.3 on p. 15]), the existence of the family $K_\mu((x, w), dw_1), x \in \mathbb{R}^d, w \in W_0^{d_1}$, satisfying (i) and (ii) follows.

To prove (iii) it suffices to show that for $t \in [0, \infty[$ and for all $A_0 \in \mathcal{B}(\mathbb{R}^d)$, $A_1 \in \mathcal{B}_t(W^d)$, $A \in \mathcal{B}_t(W_0^{d_1})$ and

$$A' := \{\pi_{r_1} - \pi_t \in B_1, \ldots, \pi_{r_k} - \pi_t \in B_k\},$$

$$t \leqslant r_1 < \ldots < r_k, B_1, \ldots, B_k \in \mathcal{B}(\mathbb{R}^{d_1}),$$

$$\int_{A_0} \int_{W_0^{d_1}} 1_{A \cap A'}(w) K_\mu((x,w), A_1) P^W(dw) \mu(dx)$$

$$= \int_{A_0} \int_{W_0^{d_1}} 1_{A \cap A'}(w) E_{\mu \otimes PW}\left(K_\mu(\cdot, A_1) | \mathcal{B}(\mathbb{R}^d) \otimes \mathcal{B}_t(W_0^{d_1})\right) P^W(dw) \mu(dx),$$

(E.1.3)

since the system of all $A \cap A'$, $A \in \mathcal{B}_t(W_0^{d_1})$, A' as above generates $\mathcal{B}(W_0^{d_1})$. But by part (ii) above, the left-hand side of (E.1.3) is equal to

$$\int 1_{A_0}(x) 1_{A \cap A'}(w) 1_{A_1}(w_1) P_\mu(dx\, dw_1\, dw)$$

$$= \int 1_{A_0}(X(0)) 1_{A_1}(X) 1_A(W) 1_{A'}(W)\, dP$$

(E.1.4)

$$= \int 1_{A_0}(X(0)) 1_{A_1}(X) 1_A(W) E_P(1_{A'}(W) | \mathcal{F}_t)\, dP.$$

But $1_{A'}(W)$ is P-independent of $\mathcal{F}_t$, since W is an $(\mathcal{F}_t)$-Wiener process on $(\Omega, \mathcal{F}, P)$, so

$$E_P(1_{A'}(W) | \mathcal{F}_t) = E_P(1_{A'}(W)).$$

Hence the right-hand side of (E.1.4) is equal to

$$P^W(A') \int 1_{A_0}(x) 1_A(w) 1_{A_1}(w_1) P_\mu(dx\, dw_1\, dw)$$

$$= P^W(A') \int_{A_0} \int_A K_\mu((x,w), A_1) P^W(dw) \mu(dx)$$

$$= P^W(A') \int_{A_0} \int_A E_{\mu \otimes PW}\left(K_\mu(\cdot, A_1) | \mathcal{B}(\mathbb{R}^d) \otimes \mathcal{B}_t(W_0^{d_1})\right)((x,w))$$

$$P^W(dw) \mu(dx)$$

$$= \int_{A_0} \int_{W_0^{d_1}} 1_{A \cap A'}(w) E_{\mu \otimes PW}\left(K_\mu(\cdot, A_1) | \mathcal{B}(\mathbb{R}^d) \otimes \mathcal{B}_t(W_0^{d_1})\right)((x,w))$$

$$P^W(dw) \mu(dx),$$

since A' is P^W-independent of $\mathcal{B}_t(W_0^{d_1})$. □

For $x \in \mathbb{R}^d$ define a measure Q_x on

$$(\mathbb{R}^d \times W^d \times W^d \times W_0^{d_1}, \mathcal{B}(\mathbb{R}^d) \otimes \mathcal{B}(W^d) \otimes \mathcal{B}(W^d) \otimes \mathcal{B}(W_0^{d_1}))$$

by

$$Q_x(A) := \int_{\mathbb{R}^d} \int_{W_0^{d_1}} \int_{W^d} \int_{W^d} 1_A(z, w_1, w_2, w)$$

$$K_\mu((z,w), dw_1) K_\mu((z,w), dw_2) P^W(dw) \delta_x(dz).$$

Define the stochastic basis

$$\tilde{\Omega} := \mathbb{R}^d \times W^d \times W^d \times W_0^{d_1}$$

$$\tilde{\mathcal{F}}^x := \overline{\mathcal{B}(\mathbb{R}^d) \otimes \mathcal{B}(W^d) \otimes \mathcal{B}(W^d) \otimes \mathcal{B}(W_0^{d_1})}^{Q_x}$$

$$\tilde{\mathcal{F}}^x_t := \bigcap_{\varepsilon>0} \sigma(\mathcal{B}(\mathbb{R}^d) \otimes \mathcal{B}_{t+\varepsilon}(W^d) \otimes \mathcal{B}_{t+\varepsilon}(W^d) \otimes \mathcal{B}_{t+\varepsilon}(W_0^{d_1}), \mathcal{N}_x),$$

where

$$\mathcal{N}_x := \{N \in \tilde{\mathcal{F}}^x | Q_x(N) = 0\},$$

and define maps

$$\Pi_0 : \tilde{\Omega} \to \mathbb{R}^d, \ (x, w_1, w_2, w) \mapsto x,$$

$$\Pi_i : \tilde{\Omega} \to W^d, \ (x, w_1, w_2, w) \mapsto w_i \in W^d, \quad i = 1, 2,$$

$$\Pi_3 : \tilde{\Omega} \to W_0^{d_1}, \ (x, w_1, w_2, w) \mapsto w \in W_0^{d_1}.$$

Then, obviously,

$$Q_x \circ \Pi_0^{-1} = \delta_x \tag{E.1.5}$$

and

$$Q_x \circ \Pi_3^{-1} = P^W (= P \circ W^{-1}). \tag{E.1.6}$$

Lemma E.1.11. *There exists $N_0 \in \mathcal{B}(\mathbb{R}^d)$ with $\mu(N_0) = 0$ such that for all $x \in N_0^c$ we have that Π_3 is an $(\tilde{\mathcal{F}}^x_t)$-Wiener process on $(\tilde{\Omega}, \tilde{\mathcal{F}}^x, Q_x)$ taking values in $\mathbb{R}^{d_1}$.*

Proof. By definition Π_3 is $(\tilde{\mathcal{F}}^x_t)$-adapted for every $x \in \mathbb{R}^d$. Furthermore, for $0 \leqslant s < t$, $y \in \mathbb{R}^d$, and $A_0 \in \mathcal{B}(\mathbb{R}^d)$, $A_i \in \mathcal{B}_s(W^d)$, $i = 1, 2$, $A_3 \in \mathcal{B}_s(W_0^{d_1})$,

$$\int_{\mathbb{R}^d} E_{Q_x}(\exp(i\langle y, \Pi_3(t) - \Pi_3(s)\rangle) 1_{A_0 \times A_1 \times A_2 \times A_3}) \mu(dx)$$

$$= \int_{\mathbb{R}^d} \int_{W_0^{d_1}} \exp(i\langle y, w(t) - w(s)\rangle) 1_{A_0}(x) 1_{A_3}(w)$$

$$\qquad K_\mu((x, w), A_1) K_\mu((x, w), A_2) P^W(dw) \mu(dx)$$

$$= \int_{W_0^{d_1}} \exp(i\langle y, w(t) - w(s)\rangle) P^W(dw) \int_{\mathbb{R}^d} Q_x(A_0 \times A_1 \times A_2 \times A_3) \mu(dx),$$

where we used Lemma E.1.10(iii) in the last step. Now the assertion follows by (E.1.6), a monotone class argument and the same reasoning as in the proof of Proposition 2.1.13. $\qquad\square$

Lemma E.1.12. *There exists $N_1 \in \mathcal{B}(\mathbb{R}^d)$, $N_0 \subset N_1$, with $\mu(N_1) = 0$ such that for all $x \in N_1^c$, (Π_1, Π_3) and (Π_2, Π_3) with stochastic basis $(\tilde{\Omega}, \tilde{\mathcal{F}}^x, Q_x, (\tilde{\mathcal{F}}^x_t))$ are (weak) solutions of (E.1.1) such that*

$$\Pi_1(0) = \Pi_2(0) = x \quad Q_x\text{-a.e.,}$$

therefore, $\Pi_1 = \Pi_2$ Q_x-a.e.

Proof. For $i = 1, 2$ consider the set $A_i \in \tilde{\mathcal{F}}^x$ defined by

$$A_i := \left\{ \Pi_i(t) - \Pi_i(0) = \int_0^t b(s, \Pi_i)\, ds + \int_0^t \sigma(s, \Pi_i)\, d\Pi_3(s) \right.$$

$$\left. \text{for all } t \in [0, \infty[\right\} \cap \{\Pi_i(0) = \Pi_0\}.$$

Define $A \in \mathcal{B}(\mathbb{R}^d) \otimes \mathcal{B}(W^d) \otimes \mathcal{B}(W_0^{d_1})$ analogously with Π_i replaced by the canonical projection from $\mathbb{R}^d \times W^d \times W_0^{d_1}$ onto the second and Π_0, Π_3 by the canonical projection onto the first and third coordinate respectively. Then by Lemma E.1.10 (ii) for $i = 1, 2$

$$\int_{\mathbb{R}^d} \int_{W_0^{d_1}} \int_{W^d} \int_{W^d} 1_{A_i}(x, w_1, w_2, w)$$

$$K_\mu((x, w), dw_1) K_\mu((x, w), dw_2) P^W(dw) \mu(dx) \tag{E.1.7}$$

$$= P_\mu(A) = P(\{(X(0), X, W) \in A\}) = 1.$$

Since all measures in the left-hand side of (E.1.7) are probability measures, it follows that for μ-a.e. $x \in \mathbb{R}^d$

$$1 = Q_x(A_i) = Q_x(A_{i,x}),$$

where for $i = 1, 2$

$$A_{i,x} := \left\{ \Pi_i(t) - x = \int_0^t b(s, \Pi_i)\, ds + \int_0^t \sigma(s, \Pi_i)\, d\Pi_3(s), \, \forall t \in [0, \infty[\right\}.$$

Hence the first assertion follows. The second then follows by the pathwise uniqueness assumption in condition (ii) of the theorem. $\qquad \square$

Lemma E.1.13. *There exists a $\overline{\mathcal{B}(\mathbb{R}^d) \otimes \mathcal{B}(W_0^{d_1})}^{\mu \otimes P^W} / \mathcal{B}(W^d)$- measurable map*

$$F_\mu : \mathbb{R}^d \times W_0^{d_1} \to W^d$$

such that

$$K_\mu((x, w), \cdot) = \delta_{F_\mu(x, w)}$$

(= Dirac measure on $\mathcal{B}(W^d)$ with mass in $F_\mu(x, w)$)

for $\mu \otimes P^W$*-a.e.* $(x, w) \in \mathbb{R}^d \times W_0^{d_1}$. *Furthermore,* F_μ *is*
$\overline{\mathcal{B}(\mathbb{R}^d) \otimes \mathcal{B}_t(W_0^{d_1})}^{\mu \otimes P^W} / \mathcal{B}_t(W^d)$*-measurable for all* $t \in [0, \infty[$, *where*
$\overline{\mathcal{B}(\mathbb{R}^d) \otimes \mathcal{B}_t(W_0^{d_1})}^{\mu \otimes P^W}$ *denotes the completion with respect to* $\mu \otimes P^W$ *in*
$\mathcal{B}(\mathbb{R}^d) \otimes \mathcal{B}_t(W_0^{d_1})$.

Proof. By Lemma E.1.12 for all $x \in N_1^c$, we have

$$1 = Q_x(\{\Pi_1 = \Pi_2\})$$
$$= \int_{W_0^{d_1}} \int_{W^d} \int_{W^d} 1_D(w_1, w_2) K_\mu((x, w), dw_1) K_\mu((x, w), dw_2) P^W(dw),$$

where $D := \{(w_1, w_1) \in W^d \times W^d | w_1 \in W^d\}$. Hence by Lemma E.1.9 there exists $N \in \mathcal{B}(\mathbb{R}^d) \otimes \mathcal{B}(W_0^{d_1})$ such that $\mu \otimes P^W(N) = 0$ and for all $(x, w) \in N^c$ there exists $F_\mu(x, w) \in W^d$ such that

$$K_\mu((x, w), dw_1) = \delta_{F_\mu(x, w)}(dw_1).$$

Set $F_\mu(x, w) := 0$, if $(x, w) \in N$. Let $A \in \mathcal{B}(W^d)$. Then

$$\{F_\mu \in A\} = (\{F_\mu \in A\} \cap N) \cup (\{K_\mu(\cdot, A) = 1\} \cap N^c)$$

and the measurability properties of F_μ follow from Lemma E.1.10. □

Having defined the mapping F_μ let us check the conditions of Definition E.1.5 and Definition E.1.6. We start with condition 2.

Lemma E.1.14. *We have*

$$X = F_\mu(X(0), W) \quad P\text{-a.e..}$$

Proof. By Lemmas E.1.10 and E.1.13 we have

$$P(\{X = F_\mu(X(0), W)\})$$
$$= \int_{\mathbb{R}^d} \int_{W_0^{d_1}} \int_{W^d} 1_{\{w_1 = F_\mu(x, w)\}}(x, w_1, w) \delta_{F_\mu(x, w)}(dw_1) P^W(dw) \mu(dx)$$
$$= 1.$$

□

Now let us check condition 1. Let W' be another $\mathbb{R}^{d_1}$-valued standard $(\mathcal{F}_t')$-Wiener process on a stochastic basis $(\Omega', \mathcal{F}', P', (\mathcal{F}_t'))$ and $\xi : \Omega' \to \mathbb{R}^d$ an $\mathcal{F}_0'/\mathcal{B}(\mathbb{R}^d)$-measurable map and $\mu := P' \circ \xi^{-1}$. Let F_μ be as above and set

$$X' := F_\mu(\xi, W').$$

Lemma E.1.15. (X', W') *is a (weak) solution to* (E.1.1) *with* $X'(0) = \xi$ P'-*a.s..*

Proof. We have

$$P'(\{\xi = X'(0)\}) = P'(\{\xi = F_\mu(\xi, W')(0)\})$$
$$= \mu \otimes P^W(\{(x, w) \in \mathbb{R}^d \times W_0^{d_1} | x = F_\mu(x, w)\})$$
$$= P(\{X(0) = F_\mu(X(0), W)(0)\}) = 1,$$

where we used Lemma E.1.14 in the last step.

To see that (X', W') is a (weak) solution we consider the set $A \in \mathcal{B}(\mathbb{R}^d) \otimes \mathcal{B}(W^d) \otimes \mathcal{B}(W_0^{d_1})$ defined in the proof of Lemma E.1.12. We have to show that

$$P'(\{(X'(0), X', W') \in A\}) = 1.$$

But since $X'(0) = \xi$ is P'-independent of W', we have

$$\int 1_A(X'(0), F_\mu(X'(0), W'), W') \, dP'$$
$$= \int_{\mathbb{R}^d} \int_{W_0^{d_1}} 1_A(x, F_\mu(x, w), w) P^W(dw) \mu(dx)$$
$$= \int_{\mathbb{R}^d} \int_{W_0^{d_1}} \int_{W^d} 1_A(x, w_1, w) \delta_{F_\mu(x,w)}(dw_1) P^W(dw) \mu(dx)$$
$$= \int 1_A(x, w_1, w) P_\mu(dx \, dw_1 \, dw)$$
$$= P(\{(X(0), X, W) \in A\}) = 1,$$

where we used E.1.10 and E.1.11 in the second to last step. $\square$

To complete the proof we still have to construct $F \in \hat{\mathcal{E}}$ and to check the adaptiveness conditions on the corresponding mappings F_μ. Below we shall apply what we have obtained above now also to δ_x replacing μ. So, for each $x \in \mathbb{R}^d$ we have a function F_{δ_x}. Now define

$$F(x, w) := F_{\delta_x}(x, w), \ x \in \mathbb{R}^d, \ w \in W_0^{d_1}. \tag{E.1.8}$$

The proof of Theorem E.1.8 is then completed by the following lemma.

Lemma E.1.16. *Let* μ *be a probability measure on* $(\mathbb{R}^d, \mathcal{B}(\mathbb{R}^d))$ *and* $F_\mu :$ $\mathbb{R}^d \times W_0^{d_1} \to W^d$ *as constructed in Lemma E.1.13. Then for* μ-*a.e.* $x \in \mathbb{R}^d$

$$F(x, \cdot) = F_\mu(x, \cdot) \quad P^W - a.e..$$

Furthermore, $F(x, \cdot)$ *is* $\overline{\mathcal{B}_t(W_0^{d_1})}^{P^W} / \mathcal{B}_t(W^d)$-*measurable for all* $x \in \mathbb{R}^d$, $t \in [0, \infty[$, *where* $\overline{\mathcal{B}_t(W_0^{d_1})}^{P^W}$ *denotes the completion of* $\mathcal{B}_t(W_0^{d_1})$ *with respect to* P^W *in* $\mathcal{B}(W_0^{d_1})$.

Proof. Let

$$\bar{\Omega} := \mathbb{R}^d \times W^d \times W_0^{d_1}$$

$$\bar{\mathcal{F}} := \mathcal{B}(\mathbb{R}^d) \otimes \mathcal{B}(W^d) \otimes \mathcal{B}(W_0^{d_1})$$

and fix $x \in \mathbb{R}^d$. Define a measure $\bar{Q}_x$ on $(\bar{\Omega}, \bar{\mathcal{F}})$ by

$$\bar{Q}_x(A) := \int_{\mathbb{R}^d} \int_{W_0^{d_1}} \int_{W^d} 1_A(z, w_1, w) K_\mu((z, w), dw_1) P^W(dw) \delta_x(dz)$$

with K_μ as in Lemma E.1.10. Consider the stochastic basis $(\bar{\Omega}, \bar{\mathcal{F}}^x, \bar{Q}_x, (\bar{\mathcal{F}}_t^x))$ where

$$\bar{\mathcal{F}}^x := \overline{\mathcal{B}(\mathbb{R}^d) \otimes \mathcal{B}(W^d) \otimes \mathcal{B}(W_0^{d_1})}^{\bar{Q}_x},$$

$$\bar{\mathcal{F}}_t^x := \bigcap_{\varepsilon > 0} \sigma(\mathcal{B}(\mathbb{R}^d) \otimes \mathcal{B}_{t+\varepsilon}(W^d) \otimes \mathcal{B}_{t+\varepsilon}(W_0^{d_1}), \bar{\mathcal{N}}_x),$$

where $\bar{\mathcal{N}}_x := \{N \in \bar{\mathcal{F}}^x | \bar{Q}_x(N) = 0\}$. As in the proof of Lemma E.1.12 one shows that (Π, Π_3) on $(\bar{\Omega}, \bar{\mathcal{F}}^x, \bar{Q}_x, (\bar{\mathcal{F}}_t^x))$ is a (weak) solution to (E.1.1) with $\Pi(0) = x$ $\bar{Q}_x$-a.e. Here

$$\Pi_0 : \mathbb{R}^d \times W^d \times W_0^{d_1} \to \mathbb{R}^d, \ (x, w_1, w) \mapsto x,$$

$$\Pi : \mathbb{R}^d \times W^d \times W_0^{d_1} \to W^d, \ (x, w_1, w) \mapsto w_1,$$

$$\Pi_3 : \mathbb{R}^d \times W^d \times W_0^{d_1} \to W_0^{d_1}, \ (x, w_1, w) \mapsto w.$$

By Lemma E.1.15 $(F_{\delta_x}(x, \Pi_3), \Pi_3)$ on the stochastic basis $(\bar{\Omega}, \bar{\mathcal{F}}^x, \bar{Q}_x, (\bar{\mathcal{F}}_t^x))$ is a (weak) solution to (E.1.1) with

$$F_{\delta_x}(x, \Pi_3)(0) = x.$$

Hence by our pathwise uniqueness assumption (ii), it follows that

$$F_{\delta_x}(x, \Pi_3) = \Pi \quad \bar{Q}_x\text{-a.s..} \tag{E.1.9}$$

Hence for all $A \in \mathcal{B}(\mathbb{R}^d) \otimes \mathcal{B}(W^d) \otimes \mathcal{B}(W_0^{d_1})$ by Lemma E.1.13 and (E.1.9)

$$\int_{\mathbb{R}^d} \int_{W^d} \int_{W_0^{d_1}} 1_A(x, w_1, w) \delta_{F_\mu(x,w)}(dw_1) P^W(dw) \mu(dx)$$

$$= \int_{\mathbb{R}^d} \bar{Q}_x(A) \mu(dx)$$

$$= \int_{\mathbb{R}^d} \int_{\bar{\Omega}} 1_A(\Pi_0, F_{\delta_x}(x, \Pi_3), \Pi_3) \, d\bar{Q}_x \mu(dx)$$

$$= \int_{\mathbb{R}^d} \int_{W_0^{d_1}} 1_A(x, F_{\delta_x}(x, w), w) P^W(dw) \mu(dx)$$

$$= \int_{\mathbb{R}^d} \int_{W_0^{d_1}} \int_{W^d} 1_A(x, w_1, w) \delta_{F_{\delta_x}(x,w)}(dw_1) P^W(dw) \mu(dx),$$

which implies the assertion.

Let $x \in \mathbb{R}^d$, $t \in [0, \infty[$, $A \in \mathcal{B}_t(W^d)$, and define

$$\bar{F}_{\delta_x} := 1_{\{x\} \times W_0^{d_1}} F_{\delta_x}.$$

Then

$$\bar{F}_{\delta_x} = F_{\delta_x} \quad \delta_x \otimes P^W - \text{a.e.},$$

hence

$$\{\bar{F}_{\delta_x} \in A\} \in \overline{\mathcal{B}(\mathbb{R}^d) \otimes \mathcal{B}_t(W_0^{d_1})}^{\delta_x \otimes P^W}. \qquad \text{(E.1.10)}$$

But

$$\{\bar{F}_{\delta_x} \in A\} = \{x\} \times \{F_{\delta_x}(x, \cdot) \in A\} \cup (\mathbb{R}^d \backslash \{x\}) \times \{0 \in A\},$$

so by (E.1.10) it follows that

$$\{F_{\delta_x}(x, \cdot) \in A\} \in \overline{\mathcal{B}_t(W_0^{d_1})}^{P^W}.$$

$\square$

F. Strong, Mild and Weak Solutions

This chapter is a short version of Chapter 2 in [FK01]. We only state the results and refer to [FK01], [DPZ92] for the proofs.

As in previous chapters let $(U, \|\,\|_U)$ and $(H, \|\,\|)$ be separable Hilbert spaces. We take $Q = I$ and fix a cylindrical Q-Wiener process $W(t)$, $t \geqslant 0$, in U on a probability space $(\Omega, \mathcal{F}, P)$ with a normal filtration $\mathcal{F}_t$, $t \geqslant 0$. Moreover, we fix $T > 0$ and consider the following type of stochastic differential equations in H:

$$
\begin{aligned}
dX(t) &= [CX(t) + F(X(t))]\, dt + B(X(t))\, dW(t), \quad t \in [0, T], \\
X(0) &= \xi,
\end{aligned}
\tag{F.0.1}
$$

where:

- $C : D(C) \to H$ is the infinitesimal generator of a C_0-semigroup $S(t)$, $t \geqslant 0$, of linear operators on H,

- $F : H \to H$ is $\mathcal{B}(H)/\mathcal{B}(H)$-measurable,

- $B : H \to L(U, H)$,

- ξ is a H-valued, $\mathcal{F}_0$-measurable random variable.

Definition F.0.1 (mild solution). An H-valued predictable process $X(t)$, $t \in [0, T]$, is called a *mild solution* of problem (F.0.1) if

$$
\begin{aligned}
X(t) = S(t)\xi &+ \int_0^t S(t - s)F(X(s))\, ds \\
&+ \int_0^t S(t - s)B(X(s))\, dW(s) \quad P\text{-a.s.}
\end{aligned}
\tag{F.0.2}
$$

for each $t \in [0, T]$. In particular, the appearing integrals have to be well-defined.

Definition F.0.2 (analytically strong solutions). A $D(C)$-valued predictable process $X(t)$, $t \in [0, T]$, (i.e. $(s, \omega) \mapsto X(s, \omega)$ is $\mathcal{P}_T/\mathcal{B}(H)$-measurable) is called an *analytically strong solution* of problem (F.0.1) if

$$
X(t) = \xi + \int_0^t CX(s) + F(X(s))\, ds + \int_0^t B(X(s))\, dW(s) \quad P\text{-a.s.}
\tag{F.0.3}
$$

for each $t \in [0, T]$. In particular, the integrals on the right-hand side have to be well-defined, that is, $CX(t)$, $F(X(t))$, $t \in [0, T]$, are P-a.s. Bochner integrable and $\mathcal{B}(X) \in \mathcal{N}_W$.

Definition F.0.3 (analytically weak solution). An H-valued predictable process $X(t)$, $t \in [0, T]$, is called an *analytically weak solution* of problem (F.0.1) if

$$
\langle X(t), \zeta \rangle = \langle \xi, \zeta \rangle + \int_0^t \langle X(s), C^* \zeta \rangle + \langle F(X(s)), \zeta \rangle \, ds
$$

$$
+ \int_0^t \langle \zeta, B(X(s)) dW(s) \rangle \quad P\text{-a.s.}
$$

(F.0.4)

for each $t \in [0, T]$ and $\zeta \in D(C^*)$. Here $(C^*, D(C^*))$ is the adjoint of $(C, D(C))$ on H.

In particular, as in Definitions F.0.2 and F.0.1, the appearing integrals have to be well-defined.

Proposition F.0.4 (analytically weak versus analytically strong solutions).

(i) *Every analytically strong solution of problem* (F.0.1) *is also an analytically weak solution.*

(ii) *Let* $X(t)$, $t \in [0, T]$, *be an analytically weak solution of problem* (F.0.1) *with values in* $D(C)$ *such that* $B(X(t))$ *takes values in* $L_2(U, H)$ *for all* $t \in [0, T]$. *Besides we assume that*

$$
P \left(\int_0^T \|CX(t)\| \, dt < \infty \right) = 1
$$

$$
P \left(\int_0^T \|F(X(t))\| \, dt < \infty \right) = 1
$$

$$
P \left(\int_0^T \|B(X(t))\|_{L_2}^2 \, dt < \infty \right) = 1.
$$

Then the process is also an analytically strong solution.

Proposition F.0.5 (analytically weak versus mild solutions).

(i) *Let* $X(t)$, $t \in [0, T]$, *be an analytically weak solution of problem* (F.0.1) *such that* $B(X(t))$ *takes values in* $L_2(U, H)$ *for all* $t \in [0, T]$. *Besides* we assume that

$$P\left(\int_0^T \|X(t)\| \, dt < \infty\right) = 1$$

$$P\left(\int_0^T \|F(X(t))\| \, dt < \infty\right) = 1$$

$$P\left(\int_0^T \|B(X(t))\|_{L_2}^2 \, dt < \infty\right) = 1.$$

Then the process is also a mild solution.

(ii) *Let* $X(t)$, $t \in [0, T]$, *be a mild solution of problem (F.0.1) such that the mappings*

$$(t, \omega) \mapsto \int_0^t S(t - s) F(X(s, \omega)) \, ds$$

$$(t, \omega) \mapsto \int_0^t S(t - s) B(X(s)) \, dW(s)(\omega)$$

have predictable versions. In addition, we require that

$$P\left(\int_0^T \|F(X(t))\| \, dt < \infty\right) = 1$$

$$\int_0^T E\left(\int_0^t \|\langle S(t - s) B(X(s)), C^* \zeta\rangle\|_{L_2(U, \mathbb{R})}^2 \, ds\right) dt < \infty$$

for all $\zeta \in D(C^*)$.
Then the process is also an analytically weak solution.

Remark F.0.6. *The precise relation of mild and analytically weak solutions with the variational solutions from Definition 4.2.1 is obviously more difficult to describe in general. We shall concentrate just on the following quite typical special case:*
Consider the situation of Subsection 4.2, but with A and B independent of t and ω. Assume that there exist a self-adjoint operator $(C, D(C))$ on H such that $-C \geqslant const. > 0$ and $F : H \to H$ $\mathcal{B}(H)/\mathcal{B}(H)$-measurable such that

$$A(x) = C(x) + F(x), \quad x \in V,$$

and

$$V := D((-C)^{\frac{1}{2}}),$$

equipped with the graph norm of $(-C)^{\frac{1}{2}}$. Then it is easy to see that C extends to a continuous linear operator form V to V^*, again denoted by C such that for $x \in V$, $y \in D(C)$

$$\tensor[_{V^*}]{\langle Cx, y\rangle}{_V} = \langle x, Cy\rangle. \tag{F.0.5}$$

Now let X be a (variational) solution in the sense of Definition 4.2.1, then it follows immediately from (F.0.5) that X is an analytically weak solution in the sense of Definition F.0.3.

Bibliography

[Alt92]   H. W. Alt, *Lineare Funktionalanalysis*, Springer-Verlag, 1992.

[AR91]   S. Albeverio and M. Röckner, *Stochastic differential equations in infinite dimensions: solutions via Dirichlet forms*, Probab. Theory Related Fields **89** (1991), no. 3, 347–386. MR MR1113223 (92k:60123)

[Aro86]   D. G. Aronson, *The porous medium equation*, Nonlinear diffusion problems (Montecatini Terme, 1985), Lecture Notes in Math., vol. 1224, Springer, Berlin, 1986, pp. 1–46. MR MR877986 (88a:35130)

[Bau01]   H. Bauer, *Measure and integration theory*, de Gruyter Studies in Mathematics, vol. 26, Walter de Gruyter & Co., Berlin, 2001.

[Coh80]   D. L. Cohn, *Measure theory*, Birkhäuser, 1980.

[Doo53]   J. L. Doob, *Stochastic processes*, John Wiley & Sons Inc., New York, 1953. MR MR0058896 (15,445b)

[DP04]   G. Da Prato, *Kolmogorov equations for stochastic PDEs*, Advanced Courses in Mathematics – CRM Barcelona, Birkhaeuser, Basel, 2004.

[DPRLRW06]   G. Da Prato, M. Röckner, B. L. Rozowskii and F. Y. Wang, *Strong solutions of stochastic generalized porous media equations: existence, uniqueness, and ergodicity*, Comm. Partial Differential Equations **31** (2006), nos 1–3, 277–291. MR MR2209754

[DPZ92]   G. Da Prato and J. Zabczyk, *Stochastic equations in infinite dimensions*, Cambridge University Press, 1992.

[DPZ96]   _____, *Ergodicity for infinite-dimensional systems*, London Mathematical Society Lecture Note Series, vol. 229, Cambridge University Press, 1996.

[FK01]   K. Frieler and C. Knoche, *Solutions of stochastic differential equations in infinite dimensional Hilbert spaces and their dependence on initial data*, Diploma Thesis, Bielefeld University, BiBoS-Preprint E02-04-083, 2001.

[GK81]   I. Gyöngy and N. V. Krylov, *On stochastic equations with respect to semimartingales. I*, Stochastics **4** (1980/81), no. 1, 1–21. MR MR587426 (82j:60104)

[GK82]   _____, *On stochastics equations with respect to semimartingales. II. Itô formula in Banach spaces*, Stochastics **6** (1981/82), nos 3–4, 153–173. MR MR665398 (84m:60070a)

[GT83]   D. Gilbarg and N. S. Trudinger, *Elliptic partial differential equations of second order*, second ed., Grundlehren der Mathematischen Wissenschaften [Fundamental Principles of Mathematical Sciences], vol. 224, Springer-Verlag, Berlin, 1983. MR MR737190 (86c:35035)

[Gyö82]   I. Gyöngy, *On stochastic equations with respect to semimartingales. III*, Stochastics **7** (1982), no. 4, 231–254. MR MR674448 (84m:60070b)

[IW81]   N. Ikeda and S. Watanabe, *Stochastic differential equations and diffusion processes*, North-Holland Mathematical Library, vol. 24, North-Holland Publishing Co., Amsterdam, 1981.

[KR79]   N. V. Krylov and B. L. Rozowskiĭ, *Stochastic evolution equations*, Current problems in mathematics, Vol. 14 (Russian), Akad. Nauk SSSR, Vsesoyuz. Inst. Nauchn. i Tekhn. Informatsii, Moscow, 1979, pp. 71–147, 256. MR MR570795 (81m:60116)

[Kry99]   N. V. Krylov, *On Kolmogorov's equations for finite-dimensional diffusions*, Stochastic PDE's and Kolmogorov equations in infinite dimensions (Cetraro, 1998), Lecture Notes in Math., vol. 1715, Springer, Berlin, 1999, pp. 1–63. MR MR1731794 (2000k:60155)

[KS88]   I. Karatzas and S. E. Shreve, *Brownian motion and stochastic calculus*, Graduate Texts in Mathematics, vol. 113, Springer-Verlag, New York, 1988. MR MR917065 (89c:60096)

[MV92]   R. Meise and D. Vogt, *Einführung in die Funktionalanalysis*, Vieweg Verlag, 1992.

[Ond04]   M. Ondreját, *Uniqueness for stochastic evolution equations in Banach spaces*, Dissertationes Math. (Rozprawy Mat.) **426** (2004), 1–63.

[Par72]   E. Pardoux, *Sur des équations aux dérivées partielles stochastiques monotones*, C. R. Acad. Sci. Paris Sér. A-B **275** (1972), A101–A103. MR MR0312572 (47 #1129)

[Par75]   _____, *Équations aux dérivées partielles stochastiques de type monotone*, Séminaire sur les Équations aux Dérivées Partielles (1974–1975), III, Exp. No. 2, Collège de France, Paris, 1975, p. 10. MR MR0651582 (58 #31406)

[Roz90]   B. Rozowskii, *Stochastic evolution systems*, Mathematics and its Applications, no. 35, Kluwer Academic Publishers Group, Dordrecht, 1990.

[RRW06]   J. Ren, M. Röckner and F. Y. Wang, *Stochastic porous media and fast diffusion equations*, Preprint, 33 pages, 2006.

[RS72]   M. Reed and B. Simon, *Methods of modern mathematical physics*, Academic Press, 1972.

[Wal86]   J. B. Walsh, *An introduction to stochastic partial differential equations*, École d'été de probabilités de Saint-Flour, XIV— 1984, Lecture Notes in Math., vol. 1180, Springer, Berlin, 1986, pp. 265–439. MR MR876085 (88a:60114)

[WW90]   H. Weizsäcker and G. Winkler, *Stochastic integrals: an introduction*, Vieweg, 1990.

[Zab98]   J. Zabczyk, *Parabolic equations on Hilbert spaces*, Stochastic PDEs and Kolmogorov Equations in Infinite Dimensions (Giuseppe Da Prato, ed.), Lecture Notes in Mathematics, Springer Verlag, 1998, pp. 117–213.

[Zei90]   E. Zeidler, *Nonlinear functional analysis and its applications. II/B*, Springer-Verlag, New York, 1990, Nonlinear monotone operators, Translated from the German by the author and Leo F. Boron. MR MR1033498 (91b:47002)

Index

Symbols

$N(m, Q)$	Gaussian measure with mean m and covariance Q, 6
$W(t)$, $t \in [0, T]$	standard Wiener process, 13
	cylindrical Wiener process, 39
$E(X\|\mathcal{G})$	conditional expectation of X given $\mathcal{G}$, 18
$\mathcal{M}_T^2(E)$	space of all continuous E-valued, square integrable martingales, 20
$\mathcal{E}$	class of all $L(U, H)$-valued elementary processes, 22
Ω_T	$[0, T] \times \Omega$, 21
dx	Lebesgue measure, 21
P_T	$dx_{\|[0,T]} \otimes P$, 21
$\mathcal{P}_T$	predictable σ-field on Ω_T, 27
$\int \Phi(s) \, dW(s)$	stochastic integral w.r.t. W, 22
$L^p(\Omega, \mathcal{F}, \mu; X)$	set of all with respect to μ p-integrable mappings from Ω to X, 106
$L^p(\Omega, \mathcal{F}, \mu)$	$L^p(\Omega, \mathcal{F}, \mu; \mathbb{R})$
L_0^p	$L^p(\Omega, \mathcal{F}_0, P; H)$
$L^p([0, T]; H)$	$L^p([0, T], \mathcal{B}([0, T]), \, dx; H)$
$L^p([0, T], \, dx)$	$L^p([0, T], \mathcal{B}([0, T]), \, dx; \mathbb{R})$
$\|\ \|_T$	L^2-norm on $L^2(\Omega_T, \mathcal{P}_T, P_T; L_2^0)$, 25
$\mathcal{N}_W^2(0, T; H)$	$L^2(\Omega_T, \mathcal{P}_T, P_T; L_2^0)$, 28
$\mathcal{N}_W^2(0, T)$	$\mathcal{N}_W^2(0, T; H)$
$\mathcal{N}_W^2$	$\mathcal{N}_W^2(0, T; H)$
$\mathcal{N}_W(0, T; H)$	space of all stochastically integrable processes, 30
$\mathcal{N}_W(0, T)$	$\mathcal{N}_W(0, T; H)$
$L(U, H)$	space of all bounded and linear operators from U to H, 109
$L(U)$	$L(U, U)$
$L_1(U, H)$	space of all nuclear operators from U to H, 109
$\operatorname{tr} Q$	trace of Q, 109
$L_2(U, H)$	space of all Hilbert–Schmidt operators from U to H, 110
$\|\ \|_{L_2}$	Hilbert–Schmidt norm, 111

A^*	adjoint operator of $A \in L(U, H)$
$Q^{\frac{1}{2}}$	square root of $Q \in L(U)$, 25
T^{-1}	(pseudo) inverse of $T \in L(U, H)$, 115
U_0	$Q^{\frac{1}{2}}(U)$, 27
L_2^0	$L_2(Q^{\frac{1}{2}}(U), H)$, 27
$\langle u, v \rangle_0$	$\langle Q^{-\frac{1}{2}}u, Q^{-\frac{1}{2}}v \rangle_U$, 27
$L(U, H)_0$	$\{T_{U_0} \mid T \in L(U, H)\}$, 27
$M(d \times d_1, \mathbb{R})$	set of all real $d \times d_1$-matrices, 43
(V, H, V^*)	Gelfand triple, 55
$C_0^\infty(\Lambda)$	set of all infinitely differentiable real-valued functions on Λ with compact support, 62
$\|\ \|_{1,p}$	norm on $C_0^\infty(\Lambda)$, 62
$H_0^{1,p}(\Lambda)$	Sobolev space, completion of $C_0^\infty(\Lambda)$ w.r.t. $\|\ \|_{1,p}$, 62
Δ_p	p-Laplacian, $p \geqslant 2$, 65
W^d	$C([0, \infty[\to \mathbb{R}^d)$, 121
W_0^d	$\{w \in W^d \mid w(0) = 0\}$, 121
$\mathcal{B}(W^d), \mathcal{B}_t(W^d)$	121
$\mathcal{A}^{d,d_1}$	121
$\hat{\mathcal{E}}$	122

Lecture Notes in Mathematics

For information about earlier volumes
please contact your bookseller or Springer
LNM Online archive: springerlink.com

Vol. 1764: A. Cannas da Silva, Lectures on Symplectic Geometry (2001)

Vol. 1765: T. Kerler, V. V. Lyubashenko, Non-Semisimple Topological Quantum Field Theories for 3-Manifolds with Corners (2001)

Vol. 1766: H. Hennion, L. Hervé, Limit Theorems for Markov Chains and Stochastic Properties of Dynamical Systems by Quasi-Compactness (2001)

Vol. 1767: J. Xiao, Holomorphic Q Classes (2001)

Vol. 1768: M. J. Pflaum, Analytic and Geometric Study of Stratified Spaces (2001)

Vol. 1769: M. Alberich-Carramiñana, Geometry of the Plane Cremona Maps (2002)

Vol. 1770: H. Gluesing-Luerssen, Linear Delay-Differential Systems with Commensurate Delays: An Algebraic Approach (2002)

Vol. 1771: M. Émery, M. Yor (Eds.), Séminaire de Probabilités 1967-1980. A Selection in Martingale Theory (2002)

Vol. 1772: F. Burstall, D. Ferus, K. Leschke, F. Pedit, U. Pinkall, Conformal Geometry of Surfaces in S^4 (2002)

Vol. 1773: Z. Arad, M. Muzychuk, Standard Integral Table Algebras Generated by a Non-real Element of Small Degree (2002)

Vol. 1774: V. Runde, Lectures on Amenability (2002)

Vol. 1775: W. H. Meeks, A. Ros, H. Rosenberg, The Global Theory of Minimal Surfaces in Flat Spaces. Martina Franca 1999. Editor: G. P. Pirola (2002)

Vol. 1776: K. Behrend, C. Gomez, V. Tarasov, G. Tian, Quantum Comohology. Cetraro 1997. Editors: P. de Bartolomeis, B. Dubrovin, C. Reina (2002)

Vol. 1777: E. García-Río, D. N. Kupeli, R. Vázquez-Lorenzo, Osserman Manifolds in Semi-Riemannian Geometry (2002)

Vol. 1778: H. Kiechle, Theory of K-Loops (2002)

Vol. 1779: I. Chueshov, Monotone Random Systems (2002)

Vol. 1780: J. H. Bruinier, Borcherds Products on O(2,1) and Chern Classes of Heegner Divisors (2002)

Vol. 1781: E. Bolthausen, E. Perkins, A. van der Vaart, Lectures on Probability Theory and Statistics. Ecole d' Eté de Probabilités de Saint-Flour XXIX-1999. Editor: P. Bernard (2002)

Vol. 1782: C.-H. Chu, A. T.-M. Lau, Harmonic Functions on Groups and Fourier Algebras (2002)

Vol. 1783: L. Grüne, Asymptotic Behavior of Dynamical and Control Systems under Perturbation and Discretization (2002)

Vol. 1784: L. H. Eliasson, S. B. Kuksin, S. Marmi, J.-C. Yoccoz, Dynamical Systems and Small Divisors. Cetraro, Italy 1998. Editors: S. Marmi, J.-C. Yoccoz (2002)

Vol. 1785: J. Arias de Reyna, Pointwise Convergence of Fourier Series (2002)

Vol. 1786: S. D. Cutkosky, Monomialization of Morphisms from 3-Folds to Surfaces (2002)

Vol. 1787: S. Caenepeel, G. Militaru, S. Zhu, Frobenius and Separable Functors for Generalized Module Categories and Nonlinear Equations (2002)

Vol. 1788: A. Vasil'ev, Moduli of Families of Curves for Conformal and Quasiconformal Mappings (2002)

Vol. 1789: Y. Sommerhäuser, Yetter-Drinfel'd Hopf algebras over groups of prime order (2002)

Vol. 1790: X. Zhan, Matrix Inequalities (2002)

Vol. 1791: M. Knebusch, D. Zhang, Manis Valuations and Prüfer Extensions I: A new Chapter in Commutative Algebra (2002)

Vol. 1792: D. D. Ang, R. Gorenflo, V. K. Le, D. D. Trong, Moment Theory and Some Inverse Problems in Potential Theory and Heat Conduction (2002)

Vol. 1793: J. Cortés Monforte, Geometric, Control and Numerical Aspects of Nonholonomic Systems (2002)

Vol. 1794: N. Pytheas Fogg, Substitution in Dynamics, Arithmetics and Combinatorics. Editors: V. Berthé, S. Ferenczi, C. Mauduit, A. Siegel (2002)

Vol. 1795: H. Li, Filtered-Graded Transfer in Using Noncommutative Gröbner Bases (2002)

Vol. 1796: J.M. Melenk, hp-Finite Element Methods for Singular Perturbations (2002)

Vol. 1797: B. Schmidt, Characters and Cyclotomic Fields in Finite Geometry (2002)

Vol. 1798: W.M. Oliva, Geometric Mechanics (2002)

Vol. 1799: H. Pajot, Analytic Capacity, Rectifiability, Menger Curvature and the Cauchy Integral (2002)

Vol. 1800: O. Gabber, L. Ramero, Almost Ring Theory (2003)

Vol. 1801: J. Azéma, M. Émery, M. Ledoux, M. Yor (Eds.), Séminaire de Probabilités XXXVI (2003)

Vol. 1802: V. Capasso, E. Merzbach, B. G. Ivanoff, M. Dozzi, R. Dalang, T. Mountford, Topics in Spatial Stochastic Processes. Martina Franca, Italy 2001. Editor: E. Merzbach (2003)

Vol. 1803: G. Dolzmann, Variational Methods for Crystalline Microstructure – Analysis and Computation (2003)

Vol. 1804: I. Cherednik, Ya. Markov, R. Howe, G. Lusztig, Iwahori-Hecke Algebras and their Representation Theory. Martina Franca, Italy 1999. Editors: V. Baldoni, D. Barbasch (2003)

Vol. 1805: F. Cao, Geometric Curve Evolution and Image Processing (2003)

Vol. 1806: H. Broer, I. Hoveijn. G. Lunther, G. Vegter, Bifurcations in Hamiltonian Systems. Computing Singularities by Gröbner Bases (2003)

Vol. 1807: V. D. Milman, G. Schechtman (Eds.), Geometric Aspects of Functional Analysis. Israel Seminar 2000-2002 (2003)

Vol. 1808: W. Schindler, Measures with Symmetry Properties (2003)

Vol. 1809: O. Steinbach, Stability Estimates for Hybrid Coupled Domain Decomposition Methods (2003)

Vol. 1810: J. Wengenroth, Derived Functors in Functional Analysis (2003)

Vol. 1811: J. Stevens, Deformations of Singularities (2003)

Vol. 1812: L. Ambrosio, K. Deckelnick, G. Dziuk, M. Mimura, V. A. Solonnikov, H. M. Soner, Mathematical Aspects of Evolving Interfaces. Madeira, Funchal, Portugal 2000. Editors: P. Colli, J. F. Rodrigues (2003)

Vol. 1813: L. Ambrosio, L. A. Caffarelli, Y. Brenier, G. Buttazzo, C. Villani, Optimal Transportation and its Applications. Martina Franca, Italy 2001. Editors: L. A. Caffarelli, S. Salsa (2003)

Vol. 1814: P. Bank, F. Baudoin, H. Föllmer, L.C.G. Rogers, M. Soner, N. Touzi, Paris-Princeton Lectures on Mathematical Finance 2002 (2003)

Vol. 1815: A. M. Vershik (Ed.), Asymptotic Combinatorics with Applications to Mathematical Physics. St. Petersburg, Russia 2001 (2003)

Vol. 1816: S. Albeverio, W. Schachermayer, M. Talagrand, Lectures on Probability Theory and Statistics. Ecole d'Eté de Probabilités de Saint-Flour XXX-2000. Editor: P. Bernard (2003)

Vol. 1817: E. Koelink, W. Van Assche (Eds.), Orthogonal Polynomials and Special Functions. Leuven 2002 (2003)

Vol. 1818: M. Bildhauer, Convex Variational Problems with Linear, nearly Linear and/or Anisotropic Growth Conditions (2003)

Vol. 1819: D. Masser, Yu. V. Nesterenko, H. P. Schlickewei, W. M. Schmidt, M. Waldschmidt, Diophantine Approximation. Cetraro, Italy 2000. Editors: F. Amoroso, U. Zannier (2003)

Vol. 1820: F. Hiai, H. Kosaki, Means of Hilbert Space Operators (2003)

Vol. 1821: S. Teufel, Adiabatic Perturbation Theory in Quantum Dynamics (2003)

Vol. 1822: S.-N. Chow, R. Conti, R. Johnson, J. Mallet-Paret, R. Nussbaum, Dynamical Systems. Cetraro, Italy 2000. Editors: J. W. Macki, P. Zecca (2003)

Vol. 1823: A. M. Anile, W. Allegretto, C. Ringhofer, Mathematical Problems in Semiconductor Physics. Cetraro, Italy 1998. Editor: A. M. Anile (2003)

Vol. 1824: J. A. Navarro González, J. B. Sancho de Salas, $\mathscr{C}^\infty$ – Differentiable Spaces (2003)

Vol. 1825: J. H. Bramble, A. Cohen, W. Dahmen, Multiscale Problems and Methods in Numerical Simulations, Martina Franca, Italy 2001. Editor: C. Canuto (2003)

Vol. 1826: K. Dohmen, Improved Bonferroni Inequalities via Abstract Tubes. Inequalities and Identities of Inclusion-Exclusion Type. VIII, 113 p, 2003.

Vol. 1827: K. M. Pilgrim, Combinations of Complex Dynamical Systems. IX, 118 p, 2003.

Vol. 1828: D. J. Green, Gröbner Bases and the Computation of Group Cohomology. XII, 138 p, 2003.

Vol. 1829: E. Altman, B. Gaujal, A. Hordijk, Discrete-Event Control of Stochastic Networks: Multimodularity and Regularity. XIV, 313 p, 2003.

Vol. 1830: M. I. Gil', Operator Functions and Localization of Spectra. XIV, 256 p, 2003.

Vol. 1831: A. Connes, J. Cuntz, E. Guentner, N. Higson, J. E. Kaminker, Noncommutative Geometry, Martina Franca, Italy 2002. Editors: S. Doplicher, L. Longo (2004)

Vol. 1832: J. Azéma, M. Émery, M. Ledoux, M. Yor (Eds.), Séminaire de Probabilités XXXVII (2003)

Vol. 1833: D.-Q. Jiang, M. Qian, M.-P. Qian, Mathematical Theory of Nonequilibrium Steady States. On the Frontier of Probability and Dynamical Systems. IX, 280 p, 2004.

Vol. 1834: Yo. Yomdin, G. Comte, Tame Geometry with Application in Smooth Analysis. VIII, 186 p, 2004.

Vol. 1835: O.T. Izhboldin, B. Kahn, N.A. Karpenko, A. Vishik, Geometric Methods in the Algebraic Theory of Quadratic Forms. Summer School, Lens, 2000. Editor: J.-P. Tignol (2004)

Vol. 1836: C. Năstăsescu, F. Van Oystaeyen, Methods of Graded Rings. XIII, 304 p, 2004.

Vol. 1837: S. Tavaré, O. Zeitouni, Lectures on Probability Theory and Statistics. Ecole d'Eté de Probabilités de Saint-Flour XXXI-2001. Editor: J. Picard (2004)

Vol. 1838: A.J. Ganesh, N.W. O'Connell, D.J. Wischik, Big Queues. XII, 254 p, 2004.

Vol. 1839: R. Gohm, Noncommutative Stationary Processes. VIII, 170 p, 2004.

Vol. 1840: B. Tsirelson, W. Werner, Lectures on Probability Theory and Statistics. Ecole d'Eté de Probabilités de Saint-Flour XXXII-2002. Editor: J. Picard (2004)

Vol. 1841: W. Reichel, Uniqueness Theorems for Variational Problems by the Method of Transformation Groups (2004)

Vol. 1842: T. Johnsen, A. L. Knutsen, K_3 Projective Models in Scrolls (2004)

Vol. 1843: B. Jefferies, Spectral Properties of Noncommuting Operators (2004)

Vol. 1844: K.F. Siburg, The Principle of Least Action in Geometry and Dynamics (2004)

Vol. 1845: Min Ho Lee, Mixed Automorphic Forms, Torus Bundles, and Jacobi Forms (2004)

Vol. 1846: H. Ammari, H. Kang, Reconstruction of Small Inhomogeneities from Boundary Measurements (2004)

Vol. 1847: T.R. Bielecki, T. Björk, M. Jeanblanc, M. Rutkowski, J.A. Scheinkman, W. Xiong, Paris-Princeton Lectures on Mathematical Finance 2003 (2004)

Vol. 1848: M. Abate, J. E. Fornaess, X. Huang, J. P. Rosay, A. Tumanov, Real Methods in Complex and CR Geometry, Martina Franca, Italy 2002. Editors: D. Zaitsev, G. Zampieri (2004)

Vol. 1849: Martin L. Brown, Heegner Modules and Elliptic Curves (2004)

Vol. 1850: V. D. Milman, G. Schechtman (Eds.), Geometric Aspects of Functional Analysis. Israel Seminar 2002-2003 (2004)

Vol. 1851: O. Catoni, Statistical Learning Theory and Stochastic Optimization (2004)

Vol. 1852: A.S. Kechris, B.D. Miller, Topics in Orbit Equivalence (2004)

Vol. 1853: Ch. Favre, M. Jonsson, The Valuative Tree (2004)

Vol. 1854: O. Saeki, Topology of Singular Fibers of Differential Maps (2004)

Vol. 1855: G. Da Prato, P.C. Kunstmann, I. Lasiecka, A. Lunardi, R. Schnaubelt, L. Weis, Functional Analytic Methods for Evolution Equations. Editors: M. Iannelli, R. Nagel, S. Piazzera (2004)

Vol. 1856: K. Back, T.R. Bielecki, C. Hipp, S. Peng, W. Schachermayer, Stochastic Methods in Finance, Bressanone/Brixen, Italy, 2003. Editors: M. Fritelli, W. Runggaldier (2004)

Vol. 1857: M. Émery, M. Ledoux, M. Yor (Eds.), Séminaire de Probabilités XXXVIII (2005)

Vol. 1858: A.S. Cherny, H.-J. Engelbert, Singular Stochastic Differential Equations (2005)

Vol. 1859: E. Letellier, Fourier Transforms of Invariant Functions on Finite Reductive Lie Algebras (2005)

Vol. 1860: A. Borisyuk, G.B. Ermentrout, A. Friedman, D. Terman, Tutorials in Mathematical Biosciences I. Mathematical Neurosciences (2005)

Vol. 1861: G. Benettin, J. Henrard, S. Kuksin, Hamiltonian Dynamics – Theory and Applications, Cetraro, Italy, 1999. Editor: A. Giorgilli (2005)

Vol. 1862: B. Helffer, F. Nier, Hypoelliptic Estimates and Spectral Theory for Fokker-Planck Operators and Witten Laplacians (2005)

Vol. 1863: H. Führ, Abstract Harmonic Analysis of Continuous Wavelet Transforms (2005)

Vol. 1864: K. Efstathiou, Metamorphoses of Hamiltonian Systems with Symmetries (2005)

Vol. 1865: D. Applebaum, B.V. R. Bhat, J. Kustermans, J. M. Lindsay, Quantum Independent Increment Processes I. From Classical Probability to Quantum Stochastic Calculus. Editors: M. Schürmann, U. Franz (2005)

Vol. 1866: O.E. Barndorff-Nielsen, U. Franz, R. Gohm, B. Kümmerer, S. Thorbjønsen, Quantum Independent Increment Processes II. Structure of Quantum Lévy Processes, Classical Probability, and Physics. Editors: M. Schürmann, U. Franz, (2005)

Vol. 1867: J. Sneyd (Ed.), Tutorials in Mathematical Biosciences II. Mathematical Modeling of Calcium Dynamics and Signal Transduction. (2005)

Recent Reprints and New Editions